INTRODUCTION TO BIOENERGETICS

INTRODUCTION TO BIOENERGETICS

By

DR. S. SUNDARA RAJAN
Principal
Reva Institute for Science and Technology Studies
R.T. Nagar
Bangalore – 560 024

ANMOL PUBLICATIONS PVT. LTD.
NEW DELHI - 110 002 (INDIA)

ANMOL PUBLICATIONS PVT. LTD.
4374/4B, Ansari Road, Daryaganj
New Delhi - 110 002
Ph.: 3261597, 3278000
Visit us at: www.anmolbooks.com

Introduction to Bioenergetics

First Edition, 2003

ISBN 81-261-1363-4

PRINTED IN INDIA

Published by J.L. Kumar for Anmol Publications Pvt. Ltd., New Delhi - 110 002 and Printed at Arora Offset Press, Delhi.

CONTENTS

Preafce

Energy and matter are the building blocks of the universe as we know it .Energy which has been defined as the capacity to do work is the basis of all changes that has occured over a period of time since the origin of Universe.

Enregy is unique in the sense, in both living and non living systems , the rules that govern its function ' (thermodynamics) ' are the same.

INTRODUCTION TO BIOENERGETICS deals with energy transformation reactions in living systems. The first chapter deals with a general survey of energy, its types, laws of thermodynamics, free energy etc. The remaining 5 chapters analyse the metabolic processes in living organisms which are the outward manifestations of energy transformations.

The book is meant to be a text book for students of life science, pharmacy, agriculture etc at the graduate and post graduate level.

I am thankful to Mr J.L Kumar, M.D of Anmol Publications, New Delhi for publishing this book.

Dr. S. Sundara Rajan
Principal

Reva Institute For Science and Technology Studies
R.T. Nagar
Bangalore - 560 024

1

INTRODUCTION

All living cells are virtual biochemical factories where hundreds of reactions take place every second involving conversion, degradation and synthesis of chemical compounds that are vital for the maintenance of life. These reactions require energy or produce energy whic his either obtained from the breakdown of chemical compounds synthesis, or in the bonds of chemical compounds. In other words, these reactions are accompanied by energy changes. Thus, *bioenergetics may be defined as the study of energy changes that accompany biochemical reactions in living systems.* Bioenergetics may also be called biochemical thermodynamics, as the energy changes follow the laws of thermodynamics precisely.

In any biological system, the rectants have a higher energy level and they move **"energy down hill"** after the reaction. The energy thus liberated is wasted as heat either partially, or wholly. The energy that is liberated as heat cannot be put into any biological function and thus is a waste. The living systems, however, have developed means and mechanisms to trap the energy and store them in the form of energy rich chemical bonds. These 'energy currencies' readily donate the energy as and when required to do so, and help perform all the activities of living beings. One such universal energy currency is a molecule of ATP (Adenosine triphosphate). Whenever energy is required for a process, ATP is hydrolysed to ATP and the energy so released is utilized for the necessary activity. Similarly, there are a number of methods in which ATP molecules are synthesized (as during respiration).

The basic rules of bioenergetics are no different from those applicable to physico - chemical energy transformations. The only difference, perhaps, is that biomolecules involved in bioenergetics are more complex and function in a highly specialized environment. In order to understand the principles involved in bioenergetics it will be necessary first to understand the physiochemical principles involved in energy transformation reactions, and later these can be applied to biological situations.

Energy and matter

The universe is composed of only two components – *energy* and *matter* which were thought to be non interconvertible and fixed in their content. Einstein ,however, has shown that matter can be coverted into energy. What is energy? Energy may be defined as the ability or capacity of a system to carry out some work. This capacity is there in all the molecules of matter to a lesser or greater extent.

Types of energy

Basically, there are two main types of energy. These are – *potential energy* and *kinetic energy.* Potential energy is bound to the molecules. It may also be called as energy at rest i.e., not working. It can be measured in terms of how much work it is capable of doing when it is released. Potential energy is of various types – chemical, light (photic), atomic and even positional (such as that of a boulder on hill top). In biological world, potential energy source is mainly the solar radiation which is trapped, transformed and stored in the form of chemical energy during photosynthesis in greeen plants. All living beings obtain their energy requirements from the green plants directly or indirectly.

Kinetic energy is the other form of energy. It can be called as *energy in motion,* or energy released. Molecules or bodies have this energy as a result of motion. The amount of kinetic energy depends upon the mass of the body and also its velocity of motion. The following equation summaries this relationship.

$$KE = \frac{mv^2}{2} \quad \text{or} \quad KE = 1/2\ mv^2 \backslash v^2 \backslash v^2$$

In the above equation KE = Kinetic energy, *m* = mass of the body/molecule, and *v*= velocity of the body/molecule.

Kinetic energy of molecules also depends on one other parameter viz., temperature. The amount of molecular kinetic energy in a system is proportional to the absolute temperature of that system and at absolute zero (-273.18^0 C), kinetic energy disappears, i.e., it is zero. For every rise of 10^{0}C in temperature, kinetic energy increases by the ratio of 10 to the absolute temperature.

Kinetic energy of molecules may be easily demonstrated through the

Browinian movement of the particles. Brownian movement, is the to and fro movement of medium. This can be clearly seen when a beam of light is made to pass through a colloidal medium. The rate of movement of particles increases with the increase in temperature. The mechanism of Brownian movement is attributed to the random motion of molecules colliding with the visible particles, and imparting to them required energy for movement.

In addition to kinetic and potential, energy may also be of two other types. Free energy and Internal energy. Free energy may be defined as the energy that is liberated or consumed in a reaction which can be used or supplied to another system. Free energy is the most important type of energy in the biological system (see later in the chapter for a detailed discussion on free energy).

Internal energy is the energy that is associated with a system based on the constitution and movement of the molecules. Internal energy is again of two types – *Internal potential energy* and *Internal kinetic energy*. The contribution of energy due to its molecular configuration is Internal potential energy, and the energy derived from the motion of its molecules is Internal Kinetic energy. The total internal energy of a system is the sum total of both the above types of energy.

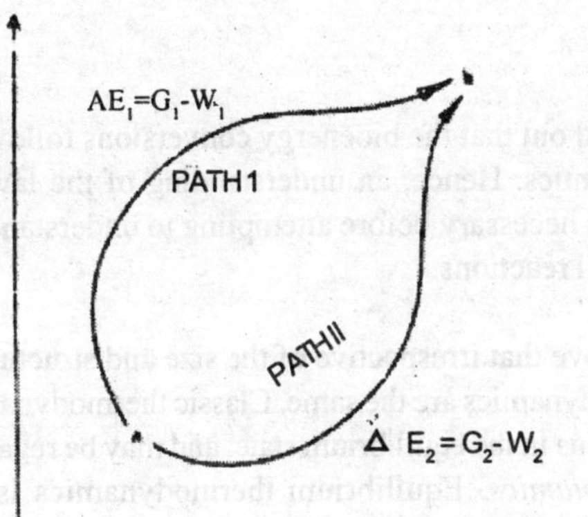

Fig. 1.1 Bioenergetics

Illustration of pathways for energy conversion of a into b

In any given substance, the internal energy is fixed under a given set of conditions. The internal energy of different substances, however, varies. Internal energy is symbolically represented by *E* or *U*.

Matter

This may be defined as something that physically occupies space and of which the entire physical universe is composed of. Originally, it was thought that matter can neither be created nor destroyed, but since the epoch making work of Albert Einstein, it has been found out that two atomic reactions *viz; nuclear fusion* and *nuclear fission* can convert matter into energy. But nuclear fusion or fission requires a very high temperature range (Nuclear fission occurs in sun and nuclear fusion is seen in atomic bombs), and they do not occur in ordinary chemical reactions.

The study of matter and its energy, conversions follow the principles of thermodynamics. In order to study the principles of thermodynamics as it relates to matter, an arbitrary position of matter, a system and its surrounding has to be imagined/considered. A system may absorb energy from the environment of liberate energy into the environment. In biological situations, *a system* may be a cell and its surroundings, or an organism and its environment or it may be an ecosystem. But the principles of thermodynamics remain the same irrespective of the size and structure of matter.

Laws of Thermodynamics

It has already been pointed out that the bioenergy conversions follow the principles of thermodynamics. Hence, an understanding of the laws of thermodynamics becomes necessary before attempting to understand the role of energy in biological reactions.

It has been mentioned above that irrespective of the size and structure of matter, the laws of thermodynamics are the same. Classic thermodynamics in reality deals with reactions in the equilibrium state, and may be regarded as *equilibrium thermodynamics*. Equilibrium thermodynamics is not concerned with the pace of the reaction, instead, it is concerned with the change in the form of energy from the initial stage (before the reaction has begun), to the final stage (when the reaction has been completed). Some of the other salient features of classic thermodynamics are –

(i) The number of steps between the initial and the final are irrelevant as long as the final state is same in different systems.

(ii) Direction of the reaction can be indicated in reversible reactions.

There are two laws of thermodynamics which are universally applicable. These are –

The first law of thermodynamics

This is known as the *Law of conservation of energy.* It states that the total content of energy is fixed and it can neither be created nor destroyed. But during reactions, energy can change form. In other words, the total energy in an isolated system will neither decrease nor increase. For instance when some amount of chemical, light or electric energy disappears (apparently), it would be transformed into equal amount of heat.

Previously, many attempts were made to disprove the principle of conservation of energy by constructing a 'perpetual motion machine', which is supposed to operate without any input of energy i.e., creating energy out of nothing. But these attempts failed.

The first law of thermodynamics may be enunciated in a detailed manner by the two following statements. These are –

(i) The disappearance of energy in one form will result in its reappearance in another form in exactly the same quality.

(ii) Energy of an isolated system will remain constant although it may be transformed from one form to another.

Elaborating the first statement, it may be said that from an isolated system when heat is given out (as in exothermic reactions), the internal energy of the reacting system decreases while that of the surrounding medium increases correspondingly. Hence, the total energy of the system (energy of the system + surrounding) remains unchanged. Alternatively, when there is energy input into the system (absorption of heat), the energy of the reacting system increases while that of the surrounding decreases. In this case also, the total energy remains unchanged.

As an illustration to elaborate the second statement, the burning of coal in a steam engine may be cited. Coal has a lot of internal energy. When coal is burnt, a lot of energy is released as heat, which is converted into mechanical energy of the engine i.e., it is allowed to boil water and release steam to rotate the wheels. In this, the chemical energy of coal gets transformed into mechanical energy of the engine.

Energy exchanges in chemical systems are studied with the help of a 'bomb calorimeter'. In the working principle, a calorimeter is nothing but a slightly modified thermos or Dewar flask. Basically, it has an insulation mechanism to isolate a system with internal energy as in a thermos flask.

A bomb calorimeter consists of an outer jacket (made up of insulating material not to allow heat to escape). This jacket surrounds a chamber which is filled with water. The centre of the calorimeter has a central chamber (combustion chamber) containing a sample which releases heat. The heat released by the sample increases the temperature of the water and its temperature is noted with the help of a thermometer. Since a calorimeter may not have a 100% insulation, corrections must be made for possible thermal leaks. With the help of this, it will be possible to show that the rise in temperature of water corresponds to the quantity of heat released by the sample.

Calculations of total energy (as per the first law)

Calculations of total energy in a reacting system can be made by finding out the quantities of the following :

(i) The work done
(ii) The heat that is exchanged, and
(iii) The energy stored in the system

Mathematically the above may be expressed as follows.

$$\Delta E = q - w$$

Where ΔE is change in the internal energy of the system, q is the heat absorbed by the system and *w* is the work done.

The values of *q* and *w* may be positive or negative, depending on the nature of reaction. If heat is absorbed, *q* has a positive value. If heat is given out, *q* has negative value. In the same way if the system performs work, *w* will be positive and if the medium performs work on the system, *w* will be negative.

The change in total energy does not depend whether *q* and *w* are negative or positive. For instance if molecules of *b, the change in energy is* ΔE_1, and the other case ΔE_2 then $\Delta E_1 = \Delta E_2$ The energy change her may also be represented as –

$$\Delta E = Ea - Eb$$

Where Ea and Eb are energy levels of molecules of *a* and *b* respectively. The first law of thermodynamics needed some modification when Albert Einstein discovered that matter is convertible into energy. Accordingly, it is now agreed that matter (mass) can be converted into energy according to the following equation.

$$E = mc^2$$

Where E = energy; *m* = mass and *c* is velocity of light in cm/sec.

The reverse of this system i.e., conversion of energy into mass is also possible. Based on these observations, the first law of thermodynamics may be restated as *matter and energy can neither be created nor destroyed but can be inter converted.*

The interconversion of matter and energy however, does not occur in biological systems.

Second law of Thermodynamics

As per the first law of thermodynamics, the total energy in a reaction remains unchanged, but the transformation of energy from one to another proceeds in such a fashion that the capacity of the total energy to do work decreases. This is because all natural processes have a tendency to move towards equilibrium. The second law is actually related to the equilibrium of the process. Stated in very simple terms, it means during every energy change the capacity of total energy to do work decreases.

The second law of thermodynamics states that spontaneous reactions have a direction for the flow of energy. Once a reaction has reached an equilibrium, the reactants cannot return spontaneously to the initial stage.

Potential energy of molecules is often referred to as energy with a high degree of orderliness, whereas the kinetic energy is said to be random or disorderly. When once potential energy changes into kinetic energy, its randomness or disorder increases. This increase in randomness is called entropy. The stage of entropy is energy in that state where energy is incapable of doing any work because of its disorderliness. In otherwords when energy is transformed from one state to another state, the entropy goes on increasing untill it reaches an equilibrium. At equilibrium, entropy is maximum.

In nature, randomness or disorder is more a rule than exception. This is the reason why orderly arrangement which is special, tends to always move towards randomness. Entropy can be zero if a system is fully orderly. For instance, at absolute zero, the entropy of a crystalline substance is zero because there is no randomness.

The second law also stipulates that in any transformation of one type of potential energy into another type, the efficiency of the process is not perfect because some quantity of potential energy gets transformed into kinetic energy. Although theoritically, total kinetic energy can be transformed into potential energy, practically it is not possible because of inefficiency of potential energy conversions. Some transformation does occur as in the case of steam turbine or internal combustion engine, but in these instances only a small fraction of the kinetic energy becomes potential mechanical energy.

Ultimately, all potential energy of the system (whether it be a molecule, body, or even the entire universe) gets converted into kinetic energy or heat and in course of time (in finite), the entire potential energy may get exhausted i.e. Total entropy.

The fundamental principles of the second law of thermodynamics also operates in a food chain, in an ecosystem where starting from producers with a high degree of order (more number of individuals), succeedingly there will be less and less consumersin number of the priinary, secondary and tertiary variety respectively. This is due to the fact that the whole of the potential energy of a producer cannot be transformed into the potential energy of a consumer. This is because of the intervening kinetic energy which introduces some amount of randomness (entropy).

Operation of the principles of second law of thermodynamics can be seen in our day to day life. For example water runs downhill and not uphill and when it has flown down its capacity to work gets decreased.

Green plants in this respect can be regarded as the most efficient mechanism to store energy for their own use and and for the use of other life forms. They trap the radiant energy and store an appreciable amount of it in the form of chemical energy. It is this potential energy which forms the fulcrum on which revolve all forms of life.

The concept of free energy

One of, the most important thermodynamic concepts that is of great relevance to biological systems is that of **free energy**. It is the energy that is liberated or consumed in a reaction and can be used or supplied to a system. It is denoted by the letter *G*. It may be regarded as a measure of the potential energy of a substance but it cannot be measured directly. However, change in free energy content that accompanies a reaction can be measured. This change is denoted by ΔG, may be defined as that quantity of total energy change in a system that is available to do the work. Change in free energy (ΔG) of a reaction is calculated by considering the difference between the sum of free energies of the reactants and products. This may be represented as follows.

$A \longrightarrow B$

In the above, *A* is reactant and *B* is in the product. Here, free energy is the maximum energy made available, as *A* is getting converted to *B*. If the free energy content of the product B *(Gb)* is less than the free energy content of the reactant A *(Ga)*, then ΔG will be negative (Here, reactant has more energy than the product). This may be shown as follows :

$\Delta G = Gb - Ga$

Here, *Ga* is greater than *Gb*, hence ΔG is negative. There is decrease in energy level as A gets converted to *B*. Similarly, when the reaction is reversible, *B* gets converted to *A*. In this reaction the level of free energy is increased (as B has less free energy).

In this case ▲G will be positive. Based on the fact whether ▲G is positive or negative (whether it gets reduced or increased), reactions are categorized into two types **endergonic** and **exergonic.**

Exergonic reactions
These reactions do not require the input of external energy. Normally, reactants have more energy than products. Products are *downhill* in energy level while reactants are *uphill* in energy level. Hence, no energy input from the medium is required to carry out the reaction. The reaction is driven by the inherent potential energy of the reactant molecules. Hence, the reaction can take place without any external energy supply. During exergonic reactions energy is released.

Endergonic reactions
Here, the reactant molecules have less free energy than the products. That is, product is *uphill,* and reactant *downhill.* Naturally, the direction of the reaction cannot be from down hill to uphill. The reactant molecules cannot negotiate on their own the uphill migration. Hence, energy from outside has to be supplied to the reactants to make them participate in the process. Here ▲G is positive *(GB–GA* and *GB* is greater than *GA,* where *GB* is free energy of reactant, and GA is free energy of the product).

The difference between endergonic and exergonic (energy requiring and energy releasing) processes can be explained with an example. Suppose a huge boulder is on top of hill, no extra effort is required to push it down hill; just a shove would do it. It has greater potential energy because of its position on hill top. When it reaches foot hill, its free energy is less because of its different position (location). This can be called exergonic process. Suppose the same boulder has to be now pushed back (from foot hill) uphill it requires an enormous effort (input of energy), which should come from outside and comparable to input of energy into a system. The boulder that is pushed uphill will again have more amount of free energy than the boulder at foot hill. This process is called endergonic, as there is energy requirement.

Photosynthesis and respiration serve as the best examples of endergonic and exergonic processes. For instant the overall formula of the processes.

1. $C_6H_{12}O_6 + 6O_2 + 6H_2O$ ——— $6CO_2 + 12H_2O$+ energy (respiration)

2. $6CO_2 + 12h_2O \xrightarrow{\text{light energy}} C_612_{12}O_6 + 6O_2\ 6H_2O$ (photosynthesis)

In the above equation for respiration, reactant (glucose) has more free energy than products. Hence, along with products energy also will be released. Reaction takes place without any energy input into the system as the reactants are *energy uphill,* and products *energy downhill.* The reaction is said to exergonic.

Equation for photosynthesis on the other hand shows that product viz., glucose has more free energy than the reactants viz CO_2 and H_2O. Hence, to drive the reaction *uphill,* there is requirement of energy input into the system which is provided by light energy. The reaction here is said to be endergonic.

Calculation of standard free energy change

In an exergonic reaction where ▲G is negative, the rate of reaction however, is determined by the surrounding conditions like temperature (heat) etc. For instance suppose 10 molecules of reactants have to become products, not all the 10 molecules may participate in the process as they may have lost some energy because of collision among molecules. Hence, only some molecules which have the required high energy participate in a reaction. This is known as *activation energy.* Activation energy for a process is the higher potential energy that is required to start a reaction. Acvitation energy may sometimes be increased depending on the environment, like a medium having higher temperature. This means then the reactant molecules have varying speeds of the reaction even in exergonic reactions depending on the medium. For example, glucose can be oxidized to CO_2 and H_2O in different surroundings. In a bomb calorimeter, oxidation takes place in a few seconds; in living organisms it takes minutes to several hours. Similarly, even in the presence of O_2 (oxidizing agent), glucose can be kept in a bottle for years without large scale oxidation. In a reaction where A and B are reactant and product respectively.

▲G = ▲G^0 + RT B/A▲

Here ▲G^0 is standard free energy. R is the gas constant and T is absolute temperature in Kelvins. This equation shows that when conditions are ideal ▲G and ▲G^0 are equal (▲G = free energy and ▲G^0 is standard free energy). *Standard free energy change may be defined as the free energy change of chemical reaction when reactants and products are in their standard state.* In otherwords, standard free energy change of a substance is that value which it can ideally release when all conditions are optimum. Standard free energy change of a process can be used to find out whether

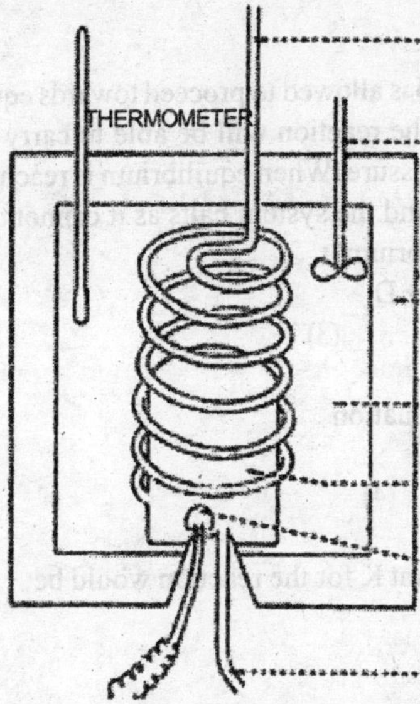

Fig. 1.2 Bioenergetics

A bomb calorimeter for determining heat of combustion of compounds a reaction is proceeding along the ideal path of not. If there is any difference between the standard free energy change and the actual free energy change that is going on in a reaction, it indicates that the conditions are not ideal necessiating corrective measures.

To summarize, the standard free energy change can be calculated as follows :

$$A + B \underline{\qquad} C + D \quad (1)$$

Where A, B, C and D are molecules of substances participating in a reaction. The free energy change ▲G, at constant temperature and pressure is given by

$$\blacktriangle G = \blacktriangle G^0 + RT \text{ in } \frac{C+D}{A+B} \quad (2)$$

In the above, ▲G = is free energy change; ▲G^0 = standard free energy change;
R=gas constant and T= absolute temperature.

In the equation (2), above ▲G^0 has a definite (fixed) value characteristic

for any given chemical reaction. Thus, we can calculate ΔG at any given temperature if ΔG^0 is known in addition to concentration of products and reactants.

If the reaction at equation (1) is allowed to proceed towards equilibrium, as the free energy decreases, the reaction will be able to carry out work at constant temperature and pressure. When equilibrium is reached, ΔG will be at its minimum or zero, and the system halts as it cannot do any more work. The equation at equilibrium is

$$O = \Delta G^0 + RT \ln \frac{C+D}{A=B} \quad (3)$$

If we rearrange the above equation

$$\Delta G^0 = - RT \ln \frac{C+D}{A+B} \quad (4)$$

Since the equilibrium constant K for the reaction would be

$$K = \frac{C+D}{A+B} \quad (5)$$

K can be substitued in the equation (4) to state as follows

$$G^0 = - RT \text{ in } K \text{ or } G^0 = 2.303 \, RT \log K$$

The above equation shows that if we can calculate the equilibrium constant for any reaction, the standard free energy change can also be calculated. In the same way, if ΔG^0 is known, ΔG is calculated either in calories/mole or joules/mole depending on whether gas constant is given in calories ($R=1.98 \, mol^{-1} \, K$-1) or joules ($R=8.31 \, J \, mol^{-1} \, k$-1)

As a conclusion to the discussion, the following summarized statements may be made about standard free energy change

(a) "The standard free energy change of a given chemical reaction is the difference between the sum of the free energies of the products and the sum of the free energies of the reactants, each reactant and product being present in its standard state" (Lehninger, 1970). The standard condition on the various components in a reaction system are –
   (i) Cencentration of 1.0m
   (ii) Temperature 25°C or 298°k
   (iii) Pressure 1.0 atm

Each compound has a particular architectural pattern that decides its inherent free energy. Hence, the standard free energy ΔG^0 of a chemical reaction may be expressed as follows

$\blacktriangle G^0 = EG^0 - EG^0$

(product) (reactant)

(b) The standard free energy change may also be expressed in terms of the equation A+B_____C+D. Here, ▲ G^0 is that quantum of free energy absorbed or lost per mole when molecules of A and B get converted into molecules of C and D, when the conditions of all the components are standard. The following table gives the free energy changes of some chemical reactions.

Standard free energy changes at pH7 and at 25°C of some chemical reactions

(Adopted from Arthur Giese, 1976)

Oxidation	**▲F**
Glucose + 60_2 ----- $6CO_2 + 6H_2O$	-6,86,000
Lactic acid + 30_2 ----- $3CO_2 + 3H_2O$	-3,26,000
Palmitic acid + $23O_2$ ----- $16CO_2 + 16H_2O$	-2,338,000
Hydrolysis	
Sucrose + H_2O ----- glucose + fructose	-5,500
Glucose 6 phosphate + H_2O ----- glucose + H_3PO_4	3,300
Glycyglocine + H_2O ----- 2 glycine	-4,600
Rearrangement	
Glucose 1, Phosphate ----- glucose 6. Phosphate	-1,745
Fructose 6, Phosphate ----- glucose 6. Phosphate	-400
Ionization	
$CH_3COOH + H_2O$ ----- $H_3O_3+ + CH_3COO -$	+6,310
Elimination	
Malate ----- fumarate + H_2O	+750

A sample calculation of ▲ G^0

In order to elucidate the calculation of ▲ G^0 from the data obtained from equilibrium constant, the reaction in which the enzyme phosphoglucomutase can be considered. This enzyme catalyses the reversible reaction of glucose 1. Phosphate to glucose 6. Phosphate. We can take 0.020m glucose 1. Phosphate add the enzyme allowing forward direction change or take 0.020m glucose 6. Phosphate and allow the reaction to go in reverse direction. In either case, at equilibrium the medium contains 0.001m glucose. 1. Phosphate and 0.019m glucose 6. Phosphate at 25°C and at photosynthesis 7.0. The equilibrium constant can be calculated as follows

$$k = \frac{\text{glucose 6. Phosphate}}{\text{glucose 1. Phosphate}} = \frac{0.019}{0.001} = 19$$

From the value of k ΔG^0 can be calculated as follows

$\Delta G^0 = RT \text{ in } K$

$= - 1.987 \times 298 \text{ In } 19$

$= - 1.987 \times 298 \times 2.303 \log 19$

$= - 1.745 \text{ cal mol}^{-1} \text{ or } - 7301 \text{ KJ mol}^{-1}$

Energy rich compounds

Living systems must get free energy from their environment in order to conduct life processes. Autotropic organisms obtain energy from sun light and their metabolism is linked to the exergonic process (respiration) to obtain energy. On the other hand, heterotrophic organisms couple their metabolism to the breakdown of complex to say that an endergonic process is always linked to an exergonic process.

In the living systems, a number of organic compounds are present which are energy rich (high energy compounds) because they undergo a large decrease in free energy change during their breakdown. These are also called energy currencies because they readily give out energy by undergoing hydrolysis. Some of these compounds are –

1. Pyrophosphate compounds
2. Acyl phosphates
3. Enolic phosphates
4. Thiol esters
5. Guanidine phosphates

Pyrophosphate compounds

Among the pyrophosphate compounds, the most important ones are ATP (Adenosine triphosphate) and ADP (Adenosine diphosphate), from the point of view of energy changes.

ATP: Historically, ATP was first discovered by C.F. Fiske and Y. Subbarow, in USA, and by K. Lohmann in Germany in 1929. After its discovery, it was thought that ATP was primarily involved in only muscle contraction reactions. Subsequent findings of Otto Warburg and Otto Meyerhof, indicated that ATP is generated from ADP during the anaerobic breakdown of glucose to lactic acid. Later, Kalckar of Denmark, and Belister of Russia, showed the generation of ATP (from ADP) during aerobic respiration also.

Hydrolysis of ATP to ADP was demonstrated by Engelhardt and Lyubjmova during energy requiring reactions.

Finally, it was Fritz Lipman in 1941, who organized all the known facts of ATP into a general hypothesis concerning energy transformation reaction in living systems. Lipman opined that – ATP functions in a cyclic manner as a carrier ATP is used for various cellular activities. When fuel molecules undergo degradation, ATP is generated by the coupled phosphorylation of ADP. The ATP so generated can donate its third phosphate group for energy requiring processes such as biosynthesis of macromolecules, transport of substances across the membrane against the concentration gradient etc. As the energy from ATP is released, it undergoes cleavage to form ADP. ADP will be rephosphorylated to ADP through the energy obtained by degradation of fuel molecules thus completing the energy cycle. The terminal (third) phosphate group is the key substance that undergoes continuous cleaving and regeneration.

Fig.1.3 Bioenergetics
space filling model of ATP

Structure of ATP, ADP and AMP

The three compounds are basically built up on the structure of ribose of deoxy ribose nucleoside. Coupled with a Phosphate group, these occur in three states – 51 monophosphate, 51 diphosphate and 51 triphosphate. Here, 51 indicates that the phosphate group(s) is attached to the 5th carbon position.

For Adenosine nucleoside also, there are three types. These are –

(a) Adenosine monophosphate (AMP)
(b) Adenosine diphosphate (ADP)
(c) Adenosine triphosphate (ATP)

The phosphate groups of these are designated as α, β, γ. AMP has only phosphate group, ADP has α, and β, while ATP has α, β, and γ phosphate groups. The γ phosphate group is energy rich and undergoes cleaving and regeneration.

The structure of ATP was first deduced by Lohmann (1930), and subsequently confirmed through total chemical synthesis by Alexander Todd *et al,* (1948). The three compounds (ATP, ADP and AMP), are of universal occurrence in all living beings.

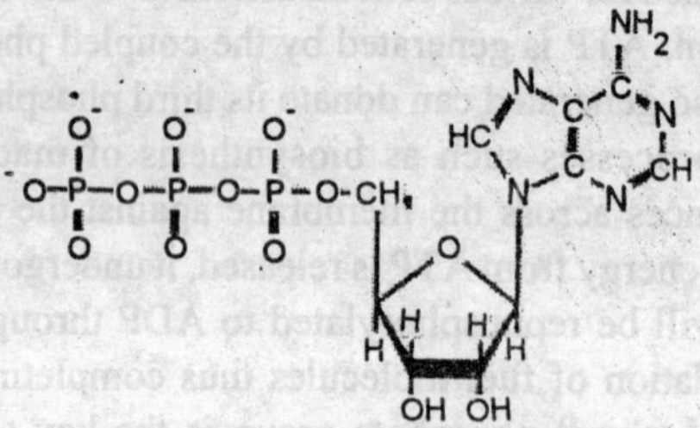

Fig.1.4 Bioenergetics
Molecular structure of ATP

The sun total of these three is constant (between 2 to 10mm) in a given species. In actively dividing cells, the amount of ATP exceeds those of ADP and AMP.

At pH 7.0 in aqueous solution both, ATP and ADP are highly charged ions. ATP has four anions while ADP has three protons. Three of the four protons of ATP get fully ionized at pH 7.0 while the fourth one will be dissociated about 75% as it has a pH of 6.95. The high concentration of closely placed negative charge around the triphosphate group in ATP has a significant bearing on its ability to carry out energy transformations.

Hydrolysis of ATP and Standard free energy

Theoretically, ▲ G^0 for ATP can be calculated by determining the equilibrium constant at pH 7.0. However there are several practical difficulties in this method as it is not easy to find out when exactly equilibrium has reached. One of the easier methods is to find out of the additive values ▲ G^0 when ATP participants are in consecutive reactions. For instance, reaction of ATP with glucose and its subsequent breakdown into glucose + phosphate will help us to calculate the total standard free energy of hydrolysis of ATP.

1. ATP + glucose ———— flucose 6. Phosphate + ADP
   Here k = 661 therefore ▲ G^0 = 4.0 cal mol

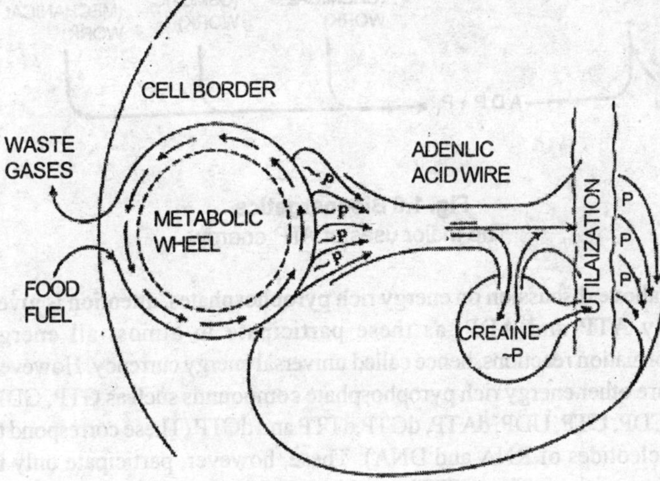

Fig. 1.5 Bioenergetics

The ATP cycle as depicted by Lipmann (1941)

2. Here k = 171 ▲ G^0 = 3.3 k cal/mol

Adding 1 and 2 of the above (-4+3.3) we get - 7.3 k cal/mol as the total standard free energy available during the hydrolysisn of ATP.

The second phosphate group of ADP also has a similar value (-7.3 k cal mol.), while the single phosphate group of AMP has G^0 value of - 3.40 k cal/ mol. This is due to the fact that the first phosphate group is linked to the nucleoside with an ester linkage, while the other two are anhydride linkages. It is well known that anhydride linkages have a great G^0 than ester linkages.

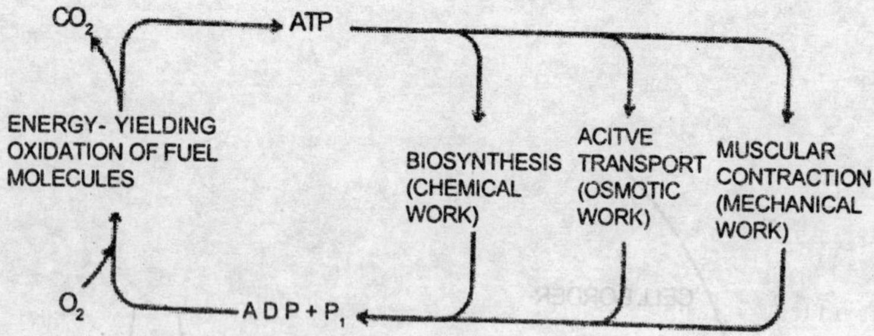

Fig. 1.6 Bioenergetics
The major uses of ATP energy

In the above discussion on energy rich pyrophosphates, attention is given to only ATP and ADP, as these participate in almost all energy transformation reactions, hence called universal energy currency. However, there are other energy rich pyrophosphate compounds such as GTP, GDP, CTP, CDP, UTP, UDP, dATP, dGTP,dTTP and dCTP (These correspond to the nucleotides of RNA and DNA). These, however, participate only in specific reactions. Thus, UTP is used in the biosynthesis of carbohydrates - polysaccharides); GTP is used in protein synthesis and CTP is utilized in lipid metabolism. These three along with ATP are involved in RNA synthesis, while dATP, dGTP, dCTP and dTTP are involved in DNA synthesis.

Another pyrophosphate compound that is energy rich is cyclic AMP (CAMP). Eventhough on hydrolysis, CAMP has a large negative free energy, it is not known to function due its high negative energy and unstable anhydride ring, instead, it acts as an allosteric effector and second messenger. Cyclic AMP is known to stimulate inactive enzymes (phosphorylase) into the active state.

2. Acyl phosphates

1.3 diphosphogylceric acid is an example of acyl phosphate. Its standard free energy on hydrolysis is – 11.8 k cal/mol. The strain of the C = 0 bond is responsible for a very high $\blacktriangle G^0$ in this kind of compounds. High energy is required (during synthesis) in these bonds to overcome the repulsive changes of carbon and phosphorous atoms and consequently on hydrolysis a large amount of energy is released.

3. Enolic phosphate

An example of enolic phosphate is phosphoenol pyruvic and PEP. This is formed during the glycolysis when glucose gets converted intp pyruvate. On hydrolysis, it has a standard free energy of - 14800 cal/mole at pH 7.0

The large negative ▲ G^0 observed on hydrolysis of this compound, is due to stabilization of enolic form of pyruvic acid in PEP by the phosphate ester group. On hydrolysis, the unstable enol will instantly isomerize into the much more stable keto structure. It has been in ▲ G of about 8000 cal/mol thus bringing the total upto - 14800 cal/mole. This tautomerization of PEP makes it one of the most 'energy rich' compounds in biological systems.

4. Thiol esters

An example of thio ester is Acetyl CoA. This is also an energy rich compound which can be utilized to generate ATP from ADP. The ▲ G^0 of this compound on hydrolysis is approximately - 75v00 cal/mole. The higher negative value for ▲ G^0 is due to the fact that thioesters do not exhibit the resonance forms that is exhibited by ordinary O_2 atom. The sulphur in thio esters do not readily release the electrons for double bond formation and hence, resonance *forms possible* for oxygen ester. Usually, oxygen es atom is knocked to the carboxyl acid carbon by means of a double bond. As the thio ester linkage is unstable (to S), the ▲ G is higher on hydrolysis.

5. Guanidine compounds

These represent a fourth type of energy rich compounds. A typical example of guanidine compound is guanidine phosphate that plays an important role in energy transfer and storage. Also known as *phosphagous,* these compounds are formed by the phosophorylation of creatine to arginine with ATP in the presence of appropriate enzyme. Hydrolysis of these compounds has a standard free energy of – 1–300 cal/mole i.e., about – 3000 more than that of ATP.

The guanidine phosphates are not inherently less stable because of bond strain as in ATP and ADP. There are no ionization or tautomerization processes which account for greater stability of reactants than their products.

In addition to the above mentioned 5 compounds which are energy rich, there are other compounds'. Some of these are, glucose 6. Phosphate, fructose 6. Phosphate etc. These have a standard free energy less than that of ATP (i.e., less than - 7300 cal/mole). The table below gives the standard free energy (on hydrolysis) of some important compounds.

Standard Free Energy on Hydrolysis of some important metabolities

	▲G' at pH 7.0 (Cal/mole
Phospheoneolypyruvate	-14,8000
Cyclic - AMP	-12,800
1,3-Disphosphyglycerate	-11,800
Phosphocreatine	-10,300
Acetyl phosphate	-10,100
S - adenosylmethionine	-10,000
Pyrophosphate	-8,000
Acetly - CoA	-7,500
ATP to ADP and Pi	-7,300
ATP to AMP and pyrophosphate	-8,600
ADP	-6,500
UDP - glucose to UDP and glucose	8,000
Glucose - 1 - Phosphate	-5,000
Fructose - 7 - Phosphate	-3,800
Glucose - 6 - Phosphate	-3,300
sn - Glycerol - 3 - Phosphate	-2,200

Coupling reactions

In biological systems, ATP is the source of energy for most of the endergonic reactions. ATP itself is synthesized in exergonic reactions by the process of phosphorylation. The endergonic and exergonic reactions are always coupled so that the energy synthesized from one (exergonic) is used to drive the other (endergonic). The net effect of this coupling is the transfer of kinetic energy from one system to the other. This coupling is mediated by a number of specific enzymes. Coupling reactions or energy coupling for efficient energy transfer is a common phenomenon. Let us consider an example in energy coupling.

The conversion of phosphoenol pyruvate to pyruvate is an exergonic process (energy releasing).

The liberated energy is used for the formation of ATP from ADP. This ATP can be hydrolysed to provide energy for other reactions. For instance, formation of glucose 6. Phosphate from glucose and phosphoric acid requires energy to the tune of – 3k cal/mole. Phosphorylation of ADP to ATP however, has free energy of –4k cal/mole. Participation of ATP in this reaction provides energy for the formation of glucose-phosphate bonds and the extra energy (of about -1k cal/ mole) is liberated into the system (waste). The above reaction may be shown as follows :

1. Glucose + Pi ———————— glucose 6, Phosphate + H_2O
(This is an endergonic reaction requiring energy of about –3k cal/mole)
2. ATP ———————— ADP + Pi

(This is an exergonic reaction releasing energy of about –4 k cal/mole)
Here reaction 1 cannot take place unless there is energy input. In order to do this reaction 1 gets coupled with 2 mediated by the enzyme hexokinase as follows :

Glucose + ATP hexokinase glucose 6. Phosphate + ADP

Energy coupling reactions produce complex compounds apparently against the second law of thermodynamics. But it has to be noted that the algebraic sum of all the components of coupled reactions is negative i.e., the sum of energy liberated is more than the sum of energy consumed (-3 k cal/mole is used while the energy liberated -4k cal/mole). In biological systems, hydrolysis of energy rich compounds is always associated with an endergonic reaction in order not to waste the energy. This requires effective energy management which is provided.

Redox potential

The ability of any particular atom, or molecule, or ion to eject an electron (getting oxidized), is measured by its oxidation – reduction potential (ability to lose an electron - oxidation, and also receive an electron - reduction). This is known as redox potential and may be defined as *"The quantitiative measure of the affinity of a compound to lose or gain electrons"*.

During oxidative phosphorylation in mitochondria ATP is produced when electron are tossed from one acceptor to another. This flow occurs according to the relative affinities of the substrate to gain or lose electrons. This is accompanied by the alternate oxidation - reduction of the system. A compound losing an electron is oxidized, while the one accepting an electron is reduced. For instance, ferric ion is oxidized to ferrous ion by losing an electron, and when ferrous ion accepts an electron, it gets reduced to ferric ion.

$$Fe^{+++} \underset{\text{reduction}}{\overset{\text{oxidation}}{\rightleftharpoons}} Fe^{++} + e$$

A redox system can be compared to a dry electric cell. In the cell, the electrons are then transferred through a wire from one electrode to another, generating electric current. In such a system the capacity to lose or gain electrons is called the electrode potential. This can be measured through a standard hydrogen electrode. The electrode potential of a reducing oxidizing system also can be measured is a similar way. The electrode potential is here called the redox potential.

2

BIOMOLECULES

Ever since its origin on this planet, life has manifested itself in myriads of forms and countless functions mainly because of the presence of unique combination of chemical compounds. These chemical compounds, the basic ingredients of which are no different from the ones present in the non living, confer upon the living the unique properties of adaptation and reproduction. The chemical compounds and their various combinations provide the living beings the unique organization and functional versatality.

The 'biomolecules' as these unique chemical compounds are popularly termed, are responsible for the origin, evolution and maintenance of life. These compounds are synthesized by the cell machinery through pathways that are unique to living beings. In this chapter we will study the basic features of three of these biomolecules–carbohydrates, fats and proteins.

CARBOHYDRATES

Carbohydrates are a group of important organic compounds present ubiquitously in plants. By being the basic respiratory substrate and in forming an integral part of the cell walls, carbohydrates are vital to the sustenance of plants, both functionally and structurally. Etymologically meaning hydrates of carbon, carbohydrates are principally composed of carbon, hydrogen and oxygen generally in the ratio of 1:2:1. Some of the carbohydrates have nitrogen and sulphur also in addition to C, H and O. Currently, however, carbohydrates are described as polyhydroxy aldehydes or polyhydroxy ketones and their derivatives.

As has been pointed out already, carbohydrates constitute the principal respiratory substrate that provides energy for all the metabolic processes in plants. They also constitute the major stored reserve in cells. In addition to these, most of the organic compounds that go to make the skeletal frame-work of the cell are composed of carbohydrates.

Classification of Carbohydrates : Carbohydrates may be classified into three principal groups viz. , Monosaccharides, Oligosaccharides and Polysaccharides.

MONOSACCHARIDES : Also known as simple sugars, monosaccharides constitute the simplest among carbohydrates. In other words hydrolysis of monosaccharides does not yield simpler carbohydrates. Monosaccharides are the basic units for the building up of oligosaccharides and polysaccharides.

Monosaccharides have carbon atoms in each molecule ranging from two to eight. This constitutes the basis for the classification of monosaccharides. Basically, all monosaccharides contain either an aldehyde group or a ketone group. Monosaccharides with aldehyde group are known as aldoses while those with a ketone group are known as ketoses.

Aldehyde Ketone

Monosaccharides are classified into the following groups based on the number of carbon atoms in each molecule.

(i) **Dioses** ($C_2H_4O_2$) : Two carbon atoms in each molecule, e.g. , Glycollic aldehyde.

(ii) **Trioses** ($C_3H_6O_3$) : These are the principle 3 carbon compounds present in the plant cells. The most important ones being glyceraldehyde and dihydroxyacetone.

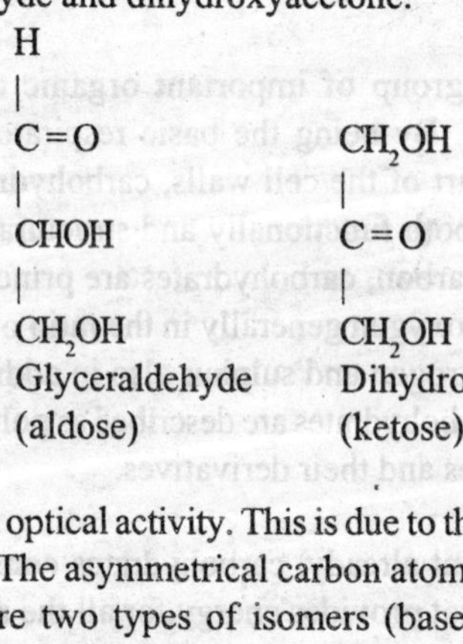

Glyceraldehyde (aldose) Dihydroxy acetone (ketose)

Triose sugars exhibit optical activity. This is due to the asymmetrical nature of the carbon atom. The asymmetrical carbon atom imparts isomerism to the trioses. There are two types of isomers (based on optical activity) dextrorotatory and laevorotatory. The former turn the plane polarized light to the right while the latter turn it to the left.

(iii) **Tetroses** ($C_4H_8O_4$) : Four carbon atoms present in each molecule, e.g. Erythrose.

(iv) **Pentoses** ($C_5H_{10}O_5$) : These are 5 carbon sugars which are generally not found in the free state in cytoplasm. They are however present as constituents of complex carbohydrates. Some of the familiar pentoses are Xylulose, Ribose, Ribulose, Arabinose etc. Xylulose and Arabinose are a part of the complex carbohydrates namely Xylans and Arabans.

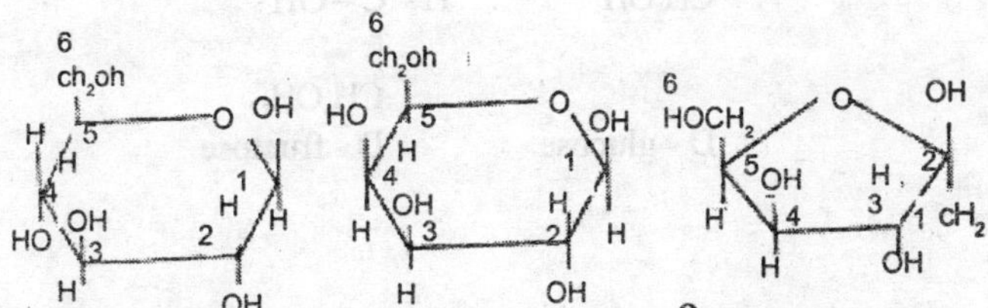

β -D-GLUCOSE β -D-GALACTOSE β -D-FRUCTOSE

Fig. 2.1 Biomolecules
Structure of glucose and fructose

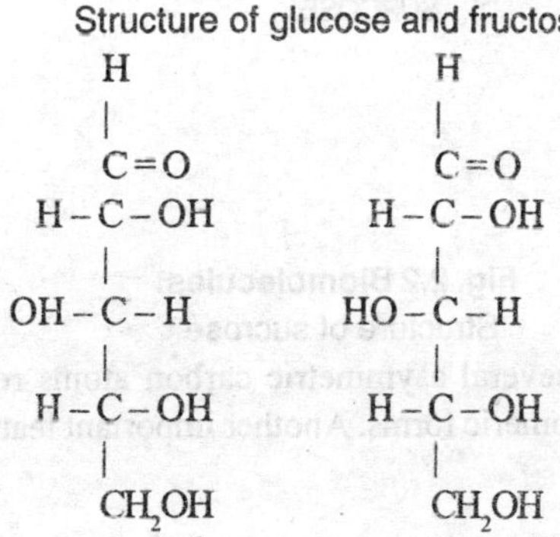

D - xylose L - arabinose

(v) **Hexoses** ($C_6H_{12}O_6$) : These are the principal sugars that play a very importat role in plant metabolism. Photosynthetic fixation of carbon results in the production of lipids via the glyoxalate cycle. Glucose and fructose are the most important hexoses found dissolved in the free form. Mannose and galactose are the other two found frequently.

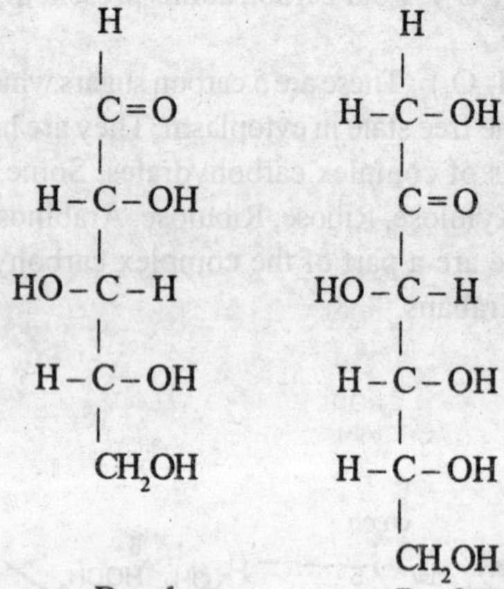

D - glucose D - fructose

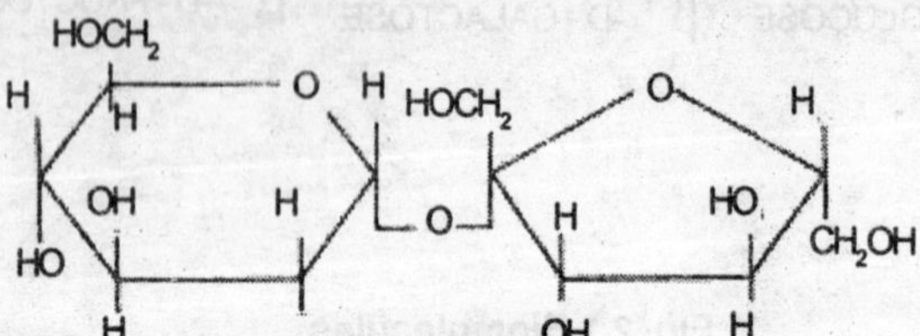

SURCOSE

Fig. 2.2 Biomolecules
Structure of sucrose

Hexose sugars have several asymmetric carbon atoms resulting in the existence of several isomeric forms. Another important feature of hexoses is their ring structure.

(vi) **Heptoses** ($C_7H_{14}O_7$) : Each molecule has 7 carbon atoms. e.g., Sedoheptulose.

OLIGOSACCHARIDES : These are condensation products of two, three or more monosaccharide units and are generally classified based on the number of monosaccharide molecules present. Thus there are disaccharides, trisaccharides, tetrasaccharides etc.

(i) **Disaccharides :** Each disaccharide consists of two units of monosaccharides. Some of the common disaccharides are Sucrose, Lactose, Maltose etc. Sucrose is the common sugar that is the cane sugar, whereas maltose is produced during seed germination by the action of amylase on starch.

(ii) **Trisaccharides :** Three units of monosaccharides condense to form one unit of a trisaccharide. E. g. Raffianose, Gentianose etc.

(iii) **Tetrasaccharides :** Four units of monosaccharides condense to form one unit of a tetrasaccharide. E. g. Stachyose.

METABOLISM OF SUCROSE

Sucrose is the principal sugar found widely distributed in plant tissues. It is made up of one unit of glucose and one unit of fructose with the elimination of one molecule, of water ($C_{12}H_{12}O_{11}$). Sucrose is a non reducing sugar even though the two units that compose it (sucrose) are reducing. Sucrose is of great importance in plant metabolism because it is in this form that carbohydrates are transported in higher plants. Radioisotopic studies have clearly shown that the labelled carbon used for photosynthesis soon finds its way in sucrose molecules.

Sucrose synthesis in plants : The work of Leloir and Cardini (1955), has shown that UDPG (Uridine diphosphoglucse) plays a key role in the addition of glucose to fructose to form a molecule of sucrose. There are at least two principal routes for sucrose synthesis. In the first pathway involving the enzyme sucrose synthetase, glucose is transferred from UDPG to fructose.

UDPG + Fructose ———— Sucrose + UDP

The second pathway is as follows.

(i) Glucose 6 phosphate ———— Glucose-l-phosphate
gluco mutase.

(ii) Glucose I phosphate combines with uridine triphosphate to produce UDPG and pyrophosphate. The enzyme Uridine diphosphoglucose pyrophosphorylase catalyses this reaction.

(iii) UDPG + fructose 6 phosphate ———— Sucrose phosphate + UDP

The enzyme sucrose phosphate synthetase catalyses this reaction.

(iv) Dephosphorylation of sucrose phosphate is brought about by the enzyme phosphatase to yield sucrose.

Sucrose degradation : Sucrose is hydrolysed by the enzyme Invertase to its unit glucose and fructose the enzymatic reaction is believed to proceed in an irreversible manner.

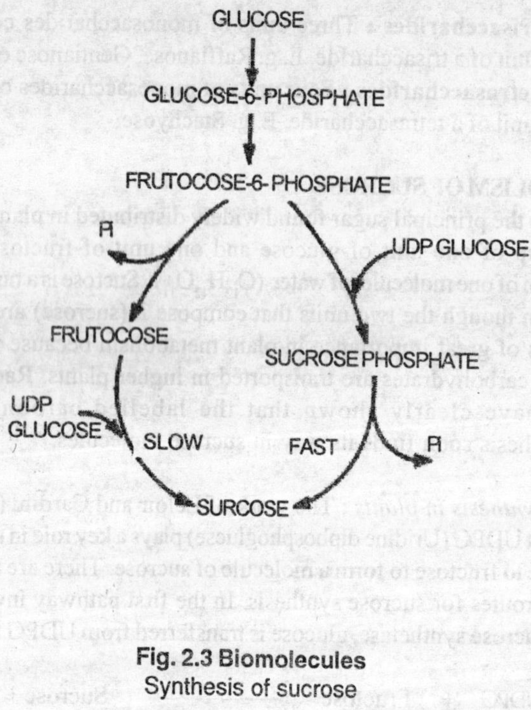

Fig. 2.3 Biomolecules
Synthesis of sucrose

POLYSACCHARIDES : These have a large number of monosaccharide units (polymer). Polysaccharides have a high molecular weight and are generally insoluble in water. Polysaccharides are found in plant cell walls. Starch and cellulose are the principal polysaccharides besides inulin, hemicellulose etc. Starch and cellulose are discussed in some detail in this book.

STARCH : This is the principle storage reserve carbohydrate. Starch is usually stored in the form of granules.

Structure of starch grains : The grains vary in shape, structure and size depending on the source. They have an average size of 1 - 50μ across, have a central point called hilum, around which are wound the chains of glucose units in a helical fashion.

Starch molecules are built up by long chains of glucose molecules joined through 1 - 4 oxygen bridges. On heating, starch yields two components viz amylose and amylopectose; the former is soluble in water while the latter is insoluble.

Each molecule of amylose consists of an unbranched helical chain of large number of glucose residues joined to each other by 1-4 oxygen bridges. Each molecule has a molecular weight from 10,000 - 10,00,000.

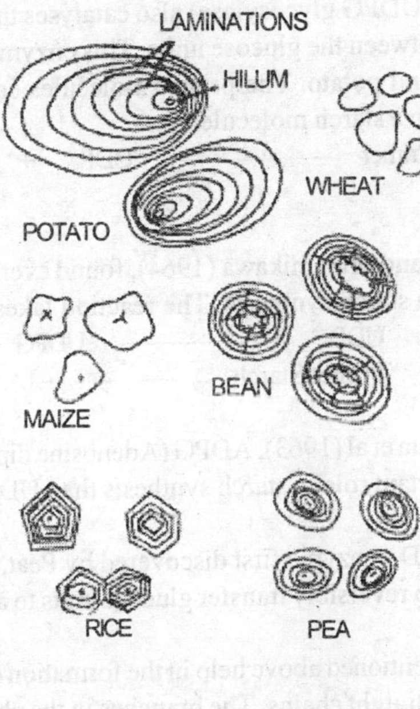

Fig. 2.4 Biomolecules
Types of starch grains

Molecules of amylopectin are made up of branched chains of glucose residues with 1-4 as well as 1-6 bridges. Branching is caused by the 1-6 oxygen linkages. Molecular weight ranges from 50,000 to 10,00,000.

Starch synthesis : Starch is synthesized either in chloroplasts or in amyloplasts. Light is not necessary for starch synthesis provided the glucose residues and the necessary enzymes are available. The following steps are involved in starch synthesis.

(i) Primer : Starch synthesis can only begin in the presence of a primer such as maltose, maltotriose etc. The primer then accepts glucose residues and forms a chain.

(ii) (Glucose 1 phosphate)n + Primer ———— Amylose + (Pi)n

The enzyme starch phosphorylase catalyzes this reaction and helps in the addition of glucose residues one by one to the non reducing end of the primer.

In addition to the above, another enzyme Uridine diphosphate transglycosylase (UDPG glycosylase) also catalyses the formation of 1-4 oxygen bridges between the glucose units. This enzyme was first isolated from bean, corn and potato. The primer molecules could be maltose or maltotriose or even a starch molecule.

UDPG + Primer ———— UDP + 1,4 - glycosyl acceptor

Akazawa, Murata and Minamikawa (1964), found even sucrose can act as a glucose donor in starch synthesis. The reaction takes place as follows.

(i) Sucrose + UDP ———— UDPG + Fructose

(ii) UDPG + Primer Starch ———— UDP

According to Murata et al (1963), ADPG (Adenosine diphosphate glucose) may play an important role in starch synthesis than UDPG.

An enzyme called D - enzyme first discovered by Peat, Whelan and Rees (1953), is known to reversibly transfer glucose units to a primer molecule.

All the enzymes mentioned above help in the formation of 1 - 4 glucosidic bridges and form straight chains. The branches in the chain formed due to 1-6 bridges are however catalyzed by an enzyme called Q - enzyme, which was first isolated from potato extracts (Baun and Gilbert). The Q - enzyme transfers small chains of glucose units on to primers at the 1 - 6 position resulting in an amylopectose molecule.

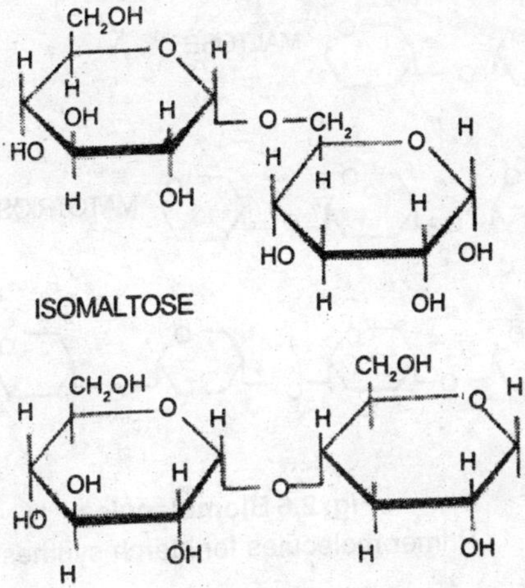

Fig. 2.5 Biomolecules
Structure of isomaltose and maltose

Starch degradation : Amylases are the principal enzymes which bring about the degradation of starch. They hydrolyse the 1-4 linkages (breaking the oxygen bridge by the addition of a water molecule) between the glucose residues and yield independent glucose moleucles. Starch phosphorylase can also reversibly degrade the starch.

CELLULOSE : This is the principal component of cell wall in most plants. It is a straight chain polymer made up of D- glucose units bound together with 1-4 bonding. Cellulose may be regarded as the most abundant natural product available from plants.

The details of cellulose synthesis and degradation are not available to the same extent as for starch. Studies conducted on Acetobacter (Celulose producing bacteria), indicate that glucose molecules have to be first phosphorylated before they are incorporated into cellulose molecule. UDPG may also play a role in cellulose synthesis.

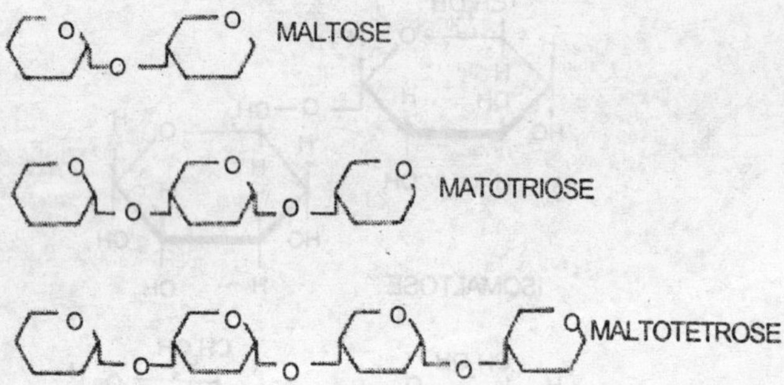

Fig. 2.6 Biomolecules
Primer molecules for starch synthesis

The enzyme complex cellulase atacks the 1-4 linkages in the chain and releases the D - glucose units as given in the figure.

FATS

Lipids constitute an important group of biomolecules playing a vital role in the storage of reserve food. Basically lipids are insoluble in water but soluble in inorganic solvents like acetone, alcohol, chloroform, benzene, ether etc. Fats are a group of simple lipids which are esters of long chain of fatty acids and a trihydric alcohol viz. , glycerol. They (fats) can undergo saponification. During esterification all the hydroxyl groups of glycerol react with fatty acids.

Like carbohydrates, fats also have C, H and O with less amount of oxygen. The following are some of the important characters of fats.

(i) All fats have glycerols, but fatty acid composition varies,
(ii) Fats are soluble in organic solvents but insoluble in water, as such they are immobile.
(iii) Oxygen content of fats is less in comparison with carbon.
(iv) Each cell synthesizes its fat requirement as they (fats) are immobile.
(v) Fats undergo saponification i.e. , when treated with alkali, they release fatty acids and glycerol.
(vi) Due to their low oxygen content, they (store and) release more energy during oxidation. On an average, a molecule of fat can yield twice as much energy as from a carbohydrate.

(vii) Fats are generally solids under room temperature, those that remain as liquids under room temperature are called oils.

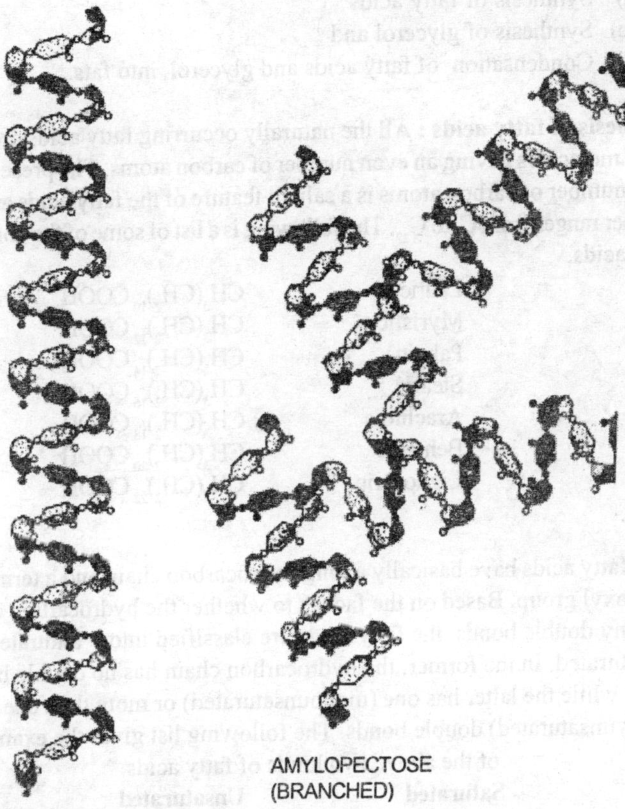

Fig. 2.7 Biomolecules
Structure of amylose and amylopectose

Distribution of fats in plants : Fats are not generally mobilised for respiration under normal conditions. They are usually the reserve food material found in leaves, stems and roots in small quantities. Major quantity of fats are however stored in seeds, either in endosperm or in cotyledons. E.g. Castor, Coconut, Peanut, Sunflower etc.

Fat metabolism : This involves two processes - synthesis and degradation. Fat synthesis involves the following three steps –

(i) Synthesis of fatty acids
(ii) Synthesis of glycerol and
(iii) Condensation of fatty acids and glycerol into fats.

Synthesis of fatty acids : All the naturally occurring fatty acids are long chain molecules having an even number of carbon atoms. The presence of even number of carbon atoms is a salient feature of the fatty acids and the number ranges from C_6 to C_{22}. The following is a list of some of the common fatty acids.

Lauric	$CH_3(CH_2)_{10}$ COOH
Myristic	$CH_3(CH_2)_{12}$ COOH
Palmitic	$CH_3(CH_2)_{14}$ COOH
Stearic	$CH_3(CH_2)_{16}$ COOH
Arachidic	$CH_3(CH_2)_{18}$ COOH
Behemic	$CH_3(CH_2)_{20}$ COOH
Lignoceric	$CH_3(CH_2)_{22}$ COOH

All fatty acids have basically a long hydrocarbon chain and a terminal carboxyl group. Based on the fact as to whether the hydrocarbon chain has any double bonds, the fatty acids are classified into – Saturated and Unsaturated. In the former, the hydrocarbon chain has no double bonds, while the latter has one (monounsaturated) or more than one (polyunsaturated) double bonds. The following list gives the examples of the above two kinds of fatty acids.

Saturated	**Unsaturated**
Lauric	Palmitoleic
Myristic	Oleic
Palmitic	Linoleic
Stearic	Linolenic
Arachidic	Archidonic

Structural arrangement of the hydrocarbon chains in palmitic (saturated) and oleic (unsaturated) is given below.

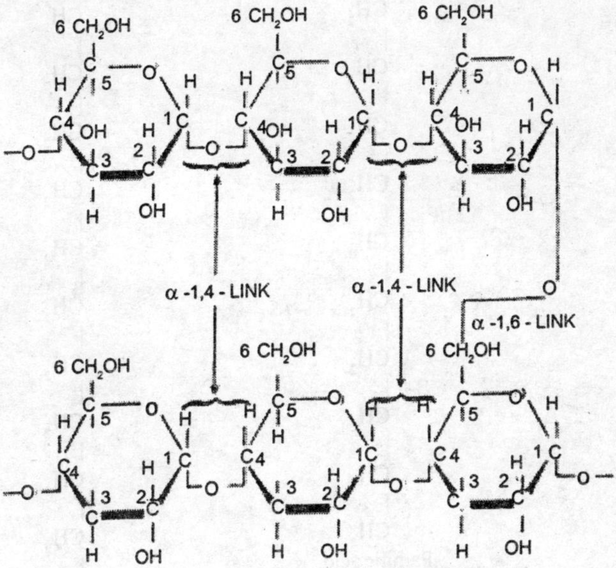

Fig. 2.8 Biomolecules
Structure of amylo pectin (detailed structure)

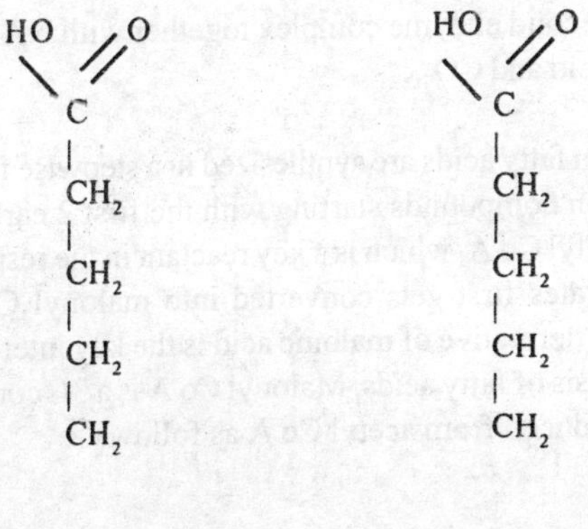

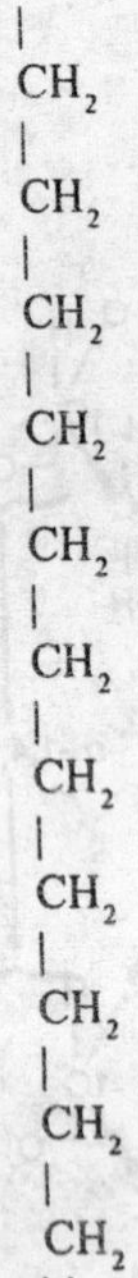

Pamitic acid

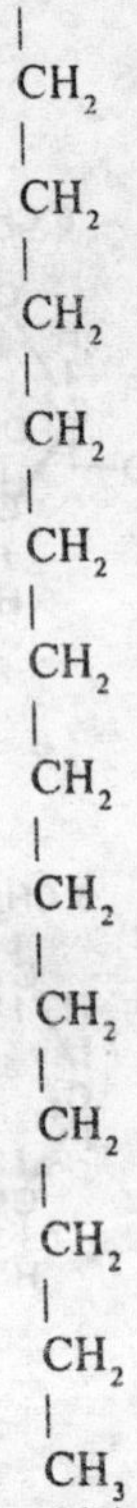

Oleic acid

Green (1960), has shown that two enzyme complexes are involved in the synthesis of fatty acids. The two enzyme complexes have five co-factors – 2ATP, Mn++, biotin, NADP and CO_2. Basically fatty acid synthesis involves two broad steps –

(i) first enzyme complex together with co-factors like ATP, Mn^{++}, biotin and CO_2 participates in the synthesis of an intermediate compound.

(ii) second enzyme complex together with NADP produces the fatty acid and CO_2.

Long chain fatty acids are synthesized in a stepwise fashion by the addition of 2 carbon compounds starting with the first 2 carbon compound acetyl Co A. Acetyl Co A, which is a key reactant in the respiratory breakdown of carbohydrates first gets converted into malonyl Co A. Malonyl Co A, which is a derivative of malonic acid is the key intermediate compound in the synthesis of fatty acids. Malonyl Co A is a 3C compound (Klein, 1957) and is produced from acetyl Co A as follows –

$$CH_3CO-S-Co\,A+HCO_3+ATP \xrightarrow[\text{Biotin}]{Mn^{++}} COOH-CH_2-CO\text{-}S\,Co\,A+ADP+iP$$

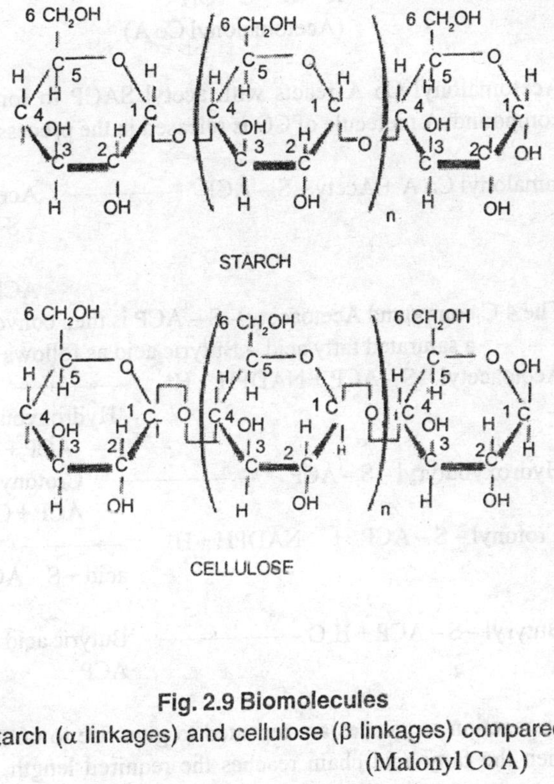

Fig. 2.9 Biomolecules
Starch (α linkages) and cellulose (β linkages) compared

(Malonyl Co A)

The synthesise of malonyl Co A is catalyzed by the enzyme acetyl Co A carboxylase with biotin as coenzyme. Further reactions involved in the synthesis of a fatty acid chain are as follows.

(i) Acetyl Co A gets attached to a protein called acetyl carrier protein (ACP) to form Acetyl Co A ACP complex.

(ii) The complex reacts with malonyl Co A to form acetomalonyl Co A, a 5 C compound.

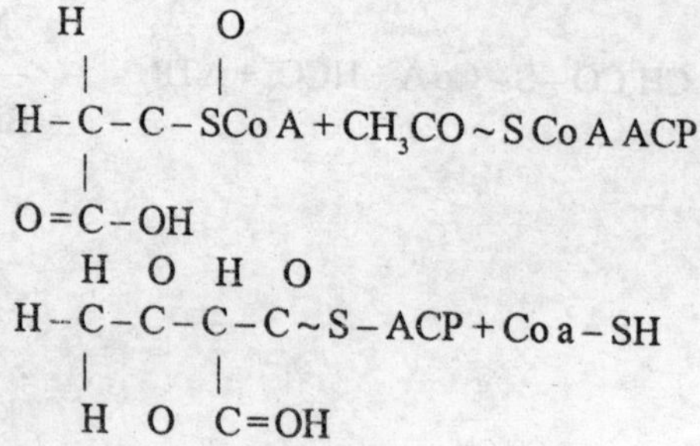

(Acetomalonyl Co A)

(iii) Acetomalonyl Co A reacts with acetyl SACP to form a 4 C compound. A molecule of CO is released in the process.

Acetomalonyl Co A + Acety – S – ACP ——— Acetoacey – S – ACP + CO_2 + ACP – SH

(iv) The 4 C compound Acetoacety – S – ACP is then converted into a saturated fatty acid – Butyric acid as follows :,

(a) Acetoacetyl – S – ACP + NADPH + H^+ ——— Hydroxybutyryl – S — ACP + $NADP^+$

(b) Hydroxybutyryl – S – ACP ——— Crotonyl – S – ACP + OH^-

(c) Crotonyl – S – ACP + NADPH + H^+ ——— Butyric acid + S – ACP + NADP +

(d) Butyryl – S – ACP + H_2O ——— Butyric acid + S – ACP

The above reaction sequences are repeated to give rise to 6C, 8C, 10C units. When the fatty acid chain reaches the required length, the acyl group is transferred to the SH group of CO A.

Synthesis of glycerol : Available evidences indicate that the synthesis of glycerol in plants involves more than one pathway. One of the well understood mechanisms, involves the 3 carbon compound dihydroxyacetone phosphate (DAP) an intermediate product in the glycolytic breakdown of carbohydrates. In the first step DAP is reduced to glycero phosphate in the presence of NADH. The reaction is catalyzed by the enzyme glycerophosphate dehydrogenase.

(i) DAP + $NADH_2$ ——— glycerophosphate + NAD^+

(ii) glycerophosphate is dephosphorylated as follows to give rise to glycerol.

$CH_2O.H_3PO_3$ | CHOH | CH_2OH + H_2O ——— CH_2OH | CHOH + H_3PO_4 | CH_2OH (Glycerol)

Condensation of fatty acids and glycerol : Fatty acids esterify the three alcoholic groups of glycerol to bring about the synthesis of fats. The three fatty acids reacting with glycerol may be same or different. The sequence of reactions is as follows:

(i) Fatty acids combine with Co A to form fatty acyl Co A complex.

(ii) Glycerol undergoes Phosphorylation to produce glycerophosphate. The reaction is catalyzed by the enzyme glycerokinase.

(iii) The acyl residues of the fatty acid acyl Co A complex are transfened to the two free hydroxyl groups of the glycerophosphate to form phosphatidic acid.

CH_2OH | CHOH | $CH_2OPO_3H_2$ (Glycerophosphate) + 2R–C(=O)–ACP (Fatty acyl CoA) → H_2COOCR | RCOOCH | $H_2COPO_3H_2$ (Phosphatidic acid) + 2ACP

(iv) In the next step, the phosphate residue is removed from the phosptatidic acid molecule to form a diglyceride.

(v) In the last step the free (unesterified) hydroxyl group of the diglyceride reacts with a fatty acyl Co A complex to form a triglyceride as follows :

RCOO CH_2 | RCOOCH | $H_2PO_3OCH_2$ (Diglyceride) + R–CO–ACP → H_2COOCR | HCOOCR | H_2COOCR (Triglyceride) + ACP

Oxidation of fats : Fats are used generally as storage reserve in plants and animals and are utilized as respiratory substrates during non availability of carbohydrates. The enzyme lipase hydrolyses fats into glycerol and fatty acids by breaking the ester bonds. This process is called saponification and is the reverse of condensation. Glycerol and fatty acids are further oxidized as follows :

Oxidation of fatty acids : There are two pathways for the oxidation of fatty acids. These are α-oxidation and β -oxidation. The type of oxidation depends on whether β -carbon atoms or α -carbon atom undergoes oxidation first.

Oxidation : This process of oxidation is known mostly in plants and is helpful in the oxidation of long chain fatty acids having more than 14 carbon atoms.

The enzyme fatty acid peroxidase brings about the oxidative decarboxyladon of fatty acids. H_2O_2 is the oxidizing agent and the fatty acids are converted into an aldehyde. The aldehyde will have one carbon less than the fatty acid.

The next step is the direct oxidation of aldehyde by an enzyme fatty aldehyde dehydrogenase. NAD is reduced to $NADH_2$ and a molecule of fatty acid with one carbon atom less is produced. The fatty acid molecule can enter the cycle once again for further decarboxylation.

Oxidation : This is the principal pathway of fatty acid oxidation. According to F. Knop (1904) who first studied this pathway, at every stage two carbon atoms are removed following the oxidation of carbon atoms. The 2 C fragments are removed in the form of acetyl Co A. The following are the sequential steps of oxidation.

(i) *Activation of fatty acids* : Fatty acids react with Co A and ATP to form Co A complex.
Fatty acid + Co A + ATP → Acyl Co – A + AMP + 2Pi.

(ii) *Dehydrogenation* : Two H Atoms between and carbon atoms are removed from the fatty acyl Co A to form trans unsaturated fatty acyl Co A. The reaction is catalyzed by the enzyme Acyl CoA dehydrogenase.
Fatty Acyl Co – A + FAD → Trans – fatty acyl Co – A+ $FADH_2$

(iii) *Hydration* : A molecule of water is added to trans – fatty Acyl Co – A to form Hydroxyacyl Co A. The enzyme enoyl hydrase catalyzes this reaction.

(iv) *Dehydrogenation* : Hydroxyacyl CoA undergoes dehydrogenation to form - Keto Acyl Co – A. The enzyme dehydrogenase catalyzes this reaction with NAD getting reduced to $NADH_2$

Hydroxy acyl Co – A + NAD^+ → Keto acyl Co – A + NADH + H^+

(v) *Thioclastic clevage* : Keto acyl Co A undergoes a clevage reaction with the elimination of a 2 carbon compound acetyl Co A. In the process, a molecule of fatty acyl Co A is produced from keto acyl Co A. Fatty acyl Co A now has two carbon atoms less than original. This re enters the cycle once again to undergo cleavage. Thioclastic cleavage is catalysed by the enzyme keto acyl thiolase.

Molecules of fatty acid which undergo oxidation become shorter by 2 carbon atoms for every step and finally get converted into 2 carbon compounds (acetyl Co A).

Acetyl Co A molecules produced as a result of oxidation enter the Krebs cycle to undergo further decarboxylation and release energy.

CONVERSION OF FATS TO CARBOHYDRATES (Glyoxylate cycle)

Acetyl Co A is the key intermediate compound between carbohydrates and fats. Oxidation of both carbohydrates and fatty acids results in the formation of acetyl CoA. While the interconversion between fatty acids and acetyl Co A is reversible, that between carbohydrates and acetyl Co A is irreversible.

(i) carbohydrates → pyruvic acid → Acetyl co A

(ii) Fatty acids → Acetyl Co A

The above reactions show that while fats can be synthesized by carbohydrates via pyruvic acid and acetycl Co A, the reverse is not possible because acetyl Co A cannot form pyruvic acid (the reaction is irreversible). In order to solve this problem, Komberg and Krebs (1957), proposed a new pathway for the synthesis of carbohydrates from fats. This pathway is known as the Glyoxylate cycle. Glyoxylate pathway was first demonstrated in microorganisms and subsequently in germinating seeds of Angiosperms when the reserve food breaks down.

Basically the Glyoxylate cycle is nothing but Krebs cycle with some modifications. The essential features of this cycle are as follows :

(i) Fatty acids get converted into fragments of acetyl Co A by oxidation, while glycerol enters into glycolysis cycle.

(ii) Acetyl Co A combines with oxaloacetic acid of the Krebs cycle to form citric acid.

(iii) Citric acid gets converted to *cis* aconitic acid and then to isocitric acid due to dehydration and rehydration.

(iv) Isocitric acid breaks up into succinic acid and Glyoxylic acid under the influence of the enzyme isocitrase.

Isocitric acid → Succinic acid + Glyoxylic acid

(v) Succinic acid continues further in the Krebs cycle by getting converted into fumaric acid, malic acid and oxaloacetic acid.

(vi) Malic acid either undergoes oxidative decarboxylation (Ochoa reaction), or reductive decarboxylation (Wood/Werkman reaction) to finally form pyruvic acid.

(a) Malic acid + NADP → Pyruvic acid + $NADPH_2$ + CO_2

(b) Malic acid + NAD → Oxaloacetic acid + $NADH_2$

(c) Oxaloacetic acid → Pyruvic acid + CO_2

Pyruvic acid gets converted into PEP (phosphoenol pyruvic''acid) which form glucose and fructose by reverse reactions of glycolysis. Glucose and fructose combine to form sucrose.

Oxidation and glyoxylate pathway occurs in certain membrane bound particles called glyoxysomes, while the conversion of succinic to oxaloacetate (Krebs Cycle) occurs in mitochondria. The site for conversion of fats to carbohydrates is therefore glyoxysomes, mitochondria and cytoplasm (from pyruvic acid to carbohydrates).

PROTEINS

Of the two important macromolecules that mastermind all the activities of the living systems, proteins are one, the nucleic acids being the other. Proteins are present all over the cell and are important both structurally and functionally.

The name protein was first suggested by Berzelius (1838). *Proteios* in Greek means first rank. Proteins are highly complex organic molecules with molecular weight ranging upto many millions. Chemically proteins are made up of C, H, O and N. In addition to these major elements, some proteins may also have sulphur, phosphorous, iron zinc, iodine etc as components. The ratio of major elements in proteins are – C = 50-55%, H = 6-8%, 0 = 20-23%, N = 15-18% and S = 0-4%.

Proteins are made up of building blocks called amino acids (i.e. on hydrolysis, proteins yield these units). The idea that proteins are built up of amino acids was first proposed by Emil Fischer.

Amino acids are carboxylic acids which have at least one COOH group and one NH, group. Other functional groups may also be present. There are atleast 22 different types of amino acids encountered in plants.
The amino acids are bound to one another in a protein chain by means of linkages or bonds called peptide linkages. Peptide linkages are formed between the COOH group of one amino acid and the NH group of the other, with the elimination of a molecule of water. Several amino acids combine together with the peptide bond and form a peptide chain. Several of these peptides constitute a polypeptide chain. The polypeptide chain folds and refolds in a characteristic fashion to form a protein molecule.

Peptides : These are made up of amino acids linked with CONH (peptide) bonds. Peptides are intermediate compounds produced during the synthesis of proteins.

Amino acids $\rightarrow$ peptides $\rightarrow$ proteins

Proteins : These are the final products in the N, metabolism, and are made up of folded peptide chains which have many additional linkages (depends upon the point of folding) in addition to the CONH bond. Proteins have a three dimensional configuration and have a very high molecular weight reaching upto 10^4 - 10^7.

Although proteins contain not more than 20 amino acids, they are of immense variety and complexity due to variations in combinations of amino acids. In general the three following aspects pertaining to the amino acids decide the type of protein. These are – the quality, quantity and sequence of amino acids in the peptide chains.

In other words, the type of amino acid, its quantity and the sequence in which these amino acids are arranged form the basis for the categorization of proteins. Besides, proteins may differ in the number of peptide chains per molecule, type of secondary linkages (SH) etc.

Structure of proteins : Modern biochemists regard four structural levels in the protein molecule which accounts for its complexity. These levels are primary, secondary, tertiary and quaternary structures, in the order of increasing complexity.

Primary structure : This is the simplest and the most basic structure and refers to the linear peptide chain. Besides the CONH bond, disulphide bonds (- S -S -) are found between or among the peptide chain, linking the sulphur containing amino acids.

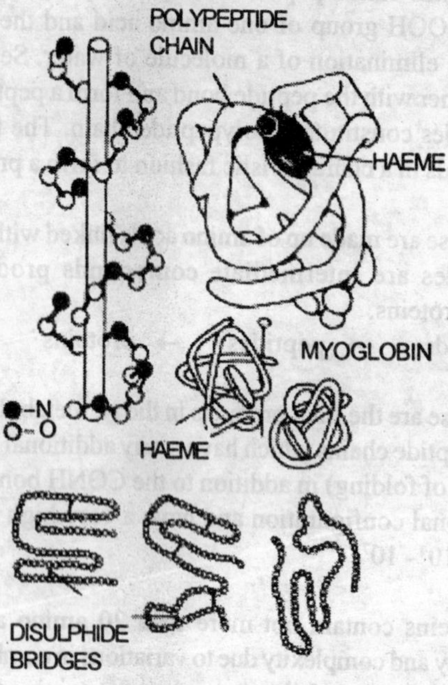

Fig. 2.10 Biomolecules
Protein structure. **A.** Polypeptide chain with helical configuration, **B.** Tertiary and quaternary structure of myoglobin, **C.** Denaturation of proteins.

Secondary structure : The primary polypeptide chain spirally coils forming the α helical chain. The characteristic bonds of the secondary structures are – hydrogen bonds between the carbonyl and amino groups of peptide links, salt linkages and Vander Walls forces which help maintain the helical structure.

Tertiary structure : Folding and superfolding of the helical chain in a three dimensional orientation constitutes the tertiary structure. Interaction between the amino acid residues situated far apart in the chain brings about the folding.

Quaternary structure: This consists of more than one identical sub units (poly peptide chain) which associate with one another to form a functional unit. Individually each subunit is non functional. Such complicated structures are usually seen in enzymes. For example, the enzyme, Phosphorylase, has two sub units, which are inactive individually, but as dimers, they are active. A protein with identical sub units in the quaternary structure is known to be homogeneous, or if the subunits are dissimilar it is known as heterogeneous. A protein with more than one sub unit (monomer) is referred to as an oligomeric protein.

Classification of Proteins : As the proteins possess, a very complex structure, their classification also poses a problem as many categorizations are possible based on their structure, function etc. The following is a generalized classification. Three major groups are recognized in proteins, based on their solubility and chemical properties.

I. Simple proteins.
II. Conjugated proteins and
III. Derived proteins.

1. SIMPLE PROTEINS : Chemically these have only amino acids. No additional compounds are present in the molecule. Simple proteins are of the .following kinds –

1. *Albumins* : These are soluble in water, dilute salts,acids and alkalies. Heat denatures the protein structure by coagulation. Some of the commonly occuring albumins are –
   Legumelin (from legume seeds),amylase (from barley),
   Leucosin (from cereals) etc.
2. *Globulins* : Less soluble in water than albumins, but soluble in less concentrated salt solutons. Heat coagulation is seen in globulins also. e.g. Legumin (pears), Tuberin(potato) etc.
3. *Glutelins* : Insoluble in water and salt solutions, but soluble in weak acids and alkalies. Glutelins are generally found in plants and are heat immune. Seed proteins found in cereals like wheat (glutenin and rice) and rice (oryzenin) are best examples of glutelins.
4. *Prolamines* : These proteins are water insoluble, but dissolve in 70 - 80% alcohol. Prolamines have a high concentration of the amino acid proline and other amides, e.g. Gliadin (Wheat), Hordein (barley), and Zein (maize)
5. *Scleroproteins* : These are fibrous proteins found as an integral part of the structural framework of plants. These proteins are insoluble in most of the solvents.

6. *Histones* : Most of the nucleoproteins, hemoglobin and globins are histones. They are insoluble in water but soluble in dilute ammonia. Histones have a high concentration of lysine or arginine.

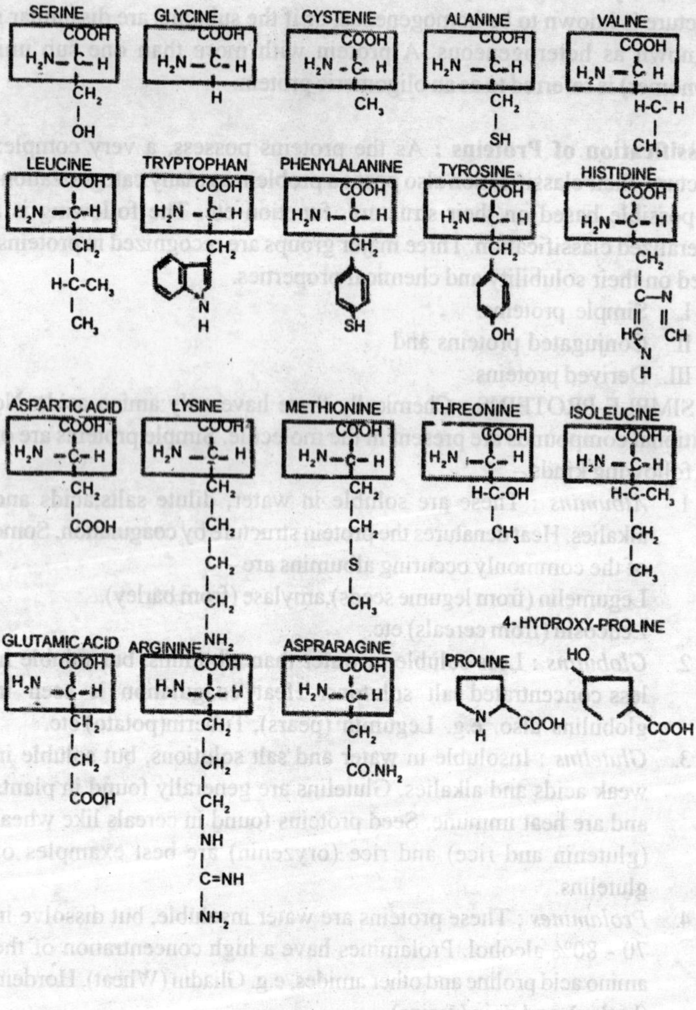

Fig. 2.11 Biomolecules
Structure of aminoacids

7. *Protamines* : These are soluble in water, dilute acids and ammonia and do not undergo heat coagulation. Basic amino acids are rich in protamines, while tryptophan and tyrosine are completely absent, e.g. Nucleoproteins of sperms in Fish.

II. CONJUGATED PROTEINS : In these proteins, non amino acid components are associated with the protein molecule. The non amino components are called prosthetic groups. There are seven major categories of conjugated proteins.

1. *Nucleoproteins* : Proteins are found in combination with nucleic acids. Hydrolysis can separate nucleic acids and simple proteins. Nucleoproteins are water soluble.
2. *Glycoproteins* : Also called mucoids, these are conjugated with carbohydrates like macromolecular polysaccharides, sugars, phosphate etc. Glycoproteins are found in cell membranes.
3. *Chromoproteins* : These constitute the pigments and are coloured, the coloured appearance is due to the presence of metals such as Cu, Fe, Mg etc. Chlorophylls, flavoproteins, carotenoids, cytochromes etc are typical examples of chromoproteins.
4. *Lipoproteins* : Proteins in conjugation with lipids are called lipoproteins. Lecithin, Cephalin etc are the lipids found in combination with proteins. Lipoproteins are an integral part of all cell membranes.
5. *Metalloproteins* : Certain types of enzymes which require metals as activators belong to this group. Many of the respiratory enzymes belong to this group.
6. *Phosphoproteins* : These are proteins with phosphoric acid as the prosthetic group. Phosphoproteins are insoluble in water, but soluble in alkalies e.g. milk and egg protein (casein and vitellin).
7. *Lecithoprotein* : Proteins with lecthin as the prosthetic group (e.g. white of egg) belong to this group.

III. DERIVED PROTEINS : These are degradation products of proteins. Primary proteins on hydrolysis with acids or alkalies yield derived proteins. There are two types of derived proteins.

(a) Primary derived proteins and
(b) Secondary derived proteins

A. **Primary derived proteins :** There are two categories of primary derived proteins. These are metaproteins and coagulated proteins.

(i) *Metaproteins* : These are insoluble in water but dissove in dilute salt solutions, acids and alkalies. Hydrolysis of natural proteins on long treatment with alkalies or acids yield metaproteins.

(ii) *Coagulated proteins* : Treatment with heat or alcohol of natural proteins yield these proteins.

B. Secondary derived proteins : Three types of secondary derived proteins have been identified. These are proteoses, peptones and peptides.

(i) *Proteoses* : Prolonged hydrolysis of metaproteins yield proteoses (eg. Albuminoses from albumin). These are water soluble but immune to heat treatment.

(ii) *Peptones* : Peptones are obtained by continued hydrolysis of proteoses. Like proteoses, these are also water soluble and heat insensitive.

(iii) *Peptides* : Prolonged hydrolysis of natural proteins yields peptides. These are also water soluble and not coagulated by heat.

Functions of proteins : Most of the vital activities of the cell are controlled by the two master molecules namely proteins and nucleic acids. Proteins are important, both from the point of view of structure and function of the cell. Structurally proteins constitute an integral part of membranes, pigments etc. Functionally in the form of enzmymes, proteins play such a vital role in cell physiology, that without proteins (enzymes) no metabolic reaction is possible. Even the synthesis of DNA is regulted by proteins. In the form of Chromoproteins (Chlorophyll, Cytochromes etc), proteins mediate all energy transformation reactions. It is because proteins are so important that the nucleic acid directly regulates the protein synthesis.

3

THE CELL

One of the marvels of nature is the creation of tiny yet potent biological compartments called cells. There is virtually no living being in the biological world which is not save for Viruses cellular. Either one, two or many millions of cells go to make the body of a living beings. A cell may be defined as the 'structural and functional unit of a living being, it is the minimal biological unit capable of maintaining and propagating itself. The study of cell comes under the perview of cytology, physiology and anatomy; while anatomy deals mainly with the structural aspects, the other two lay emphasis both on structure and function.

History

The word cell has been derived from the Latin word *cellula,* meaning a small compartment. The term in the present sense was first used by Robert Hooke (1665), in his book *Micrographic.*

There has been some mention about cell in ancient Indian literature also. Parasara (1st century BC) in his book *Vrkshayurveda* refers to invisible compartments in the leaf and names them *Rasakosha.* He further opines that each cell has a sap *(Rasa),* not visible to the naked eye *(Anavah)* and has two boundaries *(Kalavestitena),* and some of the cells have a colouring matter *(Ranjakayukta).*

The credit however of discovering the cell goes to Robert Hooke (1965), since it's difficult to establish the discovery to the ancient Indians in the absence of complete text of Parasara and the general apathy among Indians to their scientific heritage. Robert Hooke, who constructed the first compound microscope observed the sections of cork and opined that they contain honey comb like compartments. German biologists, M.J. Schleiden and T.S. Schwann(1838), established the '*cell* theory' that all organisms are made up of cells.

Dutrochet in 1824, separated the cells from *Mimosa* plant. He boiled the leaves in nitric acid to isolate the cells. Turpin and Meyers (1830), observed that every tissue is an assemblage of cells.

One of the significant discoveries of the cell came from Robert Brown (1830). He discovered the presence of a spherical body in the centre of every cell, which he named '*nucleus*'.

In the year 1835-37, Purkinje and Mohi independently discovered that protoplasm is an important constituent of every cell and it plays an important role in every cell activity, including division. Golgi (1838), discovered the golgi apparatus. Balbiani (1881), discovered chromosomes in the salivary glands of *Chironomus.* At about the same time, Flemming (1882) studied cell division in detail and gave the name *Mitosis.*

Endoplasmic reticulum was discovered by Porter in 1945. Benda gave the name mitochondria to organelles originally discovered by Hemming. Lysosomes were discovered in 1955 by de Duve.

Shape of the Cell

The shape of the cell may be variable, as in the case of *Amoeba,* or fixed. When the shape is fixed or permanent, all conceivable shapes are found-- they may be spherical, rectangular, flattened, polygonal, oval, triangular, cone like, columnar etc. As in *Acetabularia,* the body of the plant made of a single cell is differentiated into a rhizoid like base, and long and slender stalk and an umbrella like cap. Certain cells, like the Ascospores of *Claviceps* are *acicular* are needle like.

Size

There is a great range of variation among cells in size also. Some of them are not visible to the naked eye, some are barely visible, while some are macroscopic The smallest cell size can be encountered in *Coccus* bacteria (0.2 to 0.5 m), while the largest size of cell is seen in an Ostrich egg (nearly 15 cms). The table given below summarises the variations in cell size.

Table 1 (Adopted from U. Sinha and S. Sinha, 1976)

Cell type	Size
Fibre cell of *Linum* (length)	550 mm
Egg of ostrich	170 x 35 mm
Egg of Hen	60 x 45 mm
Human egg	0.1 mm = 100 µm
Amoeba	100 µ
Sea urchin egg	70 µ
RBC	7 µ
Typhoid bacillus	2.4 x 0.5 µ
E. Coli	1.5 x 0.7 µ
Diplococcus pneumonieae	200 x 100 nm
*Influenze virus	100 nm
*TMV	300 x 15 nm
*T_3 bacteriophage	45 nm,

(Even though Viruses cannot be regarded as cells, they have been mentioned as they are independent entities).

The factors governing the cell's size are – a) ratio between the volume of nucleus and volume of cytoplasm, b) ratio of cell surface to cell volume and c) the rate of metabolism.

Gross Morphology of the cell

A generalised plant cell has an outermost envelope called the cell wall. This is absent in animal cells. Internal to this is the plasma membrane. This encloses the nucleus and other cytoplasmic inclusions suspended in cytoplasm. The chief cytoplasmic inclusions are –Ribosomes, Lysosomes, Mitochondria, Plastids, Golgi complex, Endoplasmic reticulum, Vacuole and non living inclusions like crystals, raphides etc.

Prokaryotic and Eukaryotic cells

In primitive organisms like certain bacteria, blue green algae etc., the nucleus is not properly organised, hence such cells are called prokaryotic, while in evolved organisms, the nucleus is organised. Such cells are called eukaryotic. The following are some of the fundamental differences between eukaryotic and prokaryotic cells.

1. **Nucleus :** Nuclear membrane is absent in prokaryotic cells, while eukaryotic cells have a definite bounding membrane. The nuclear material

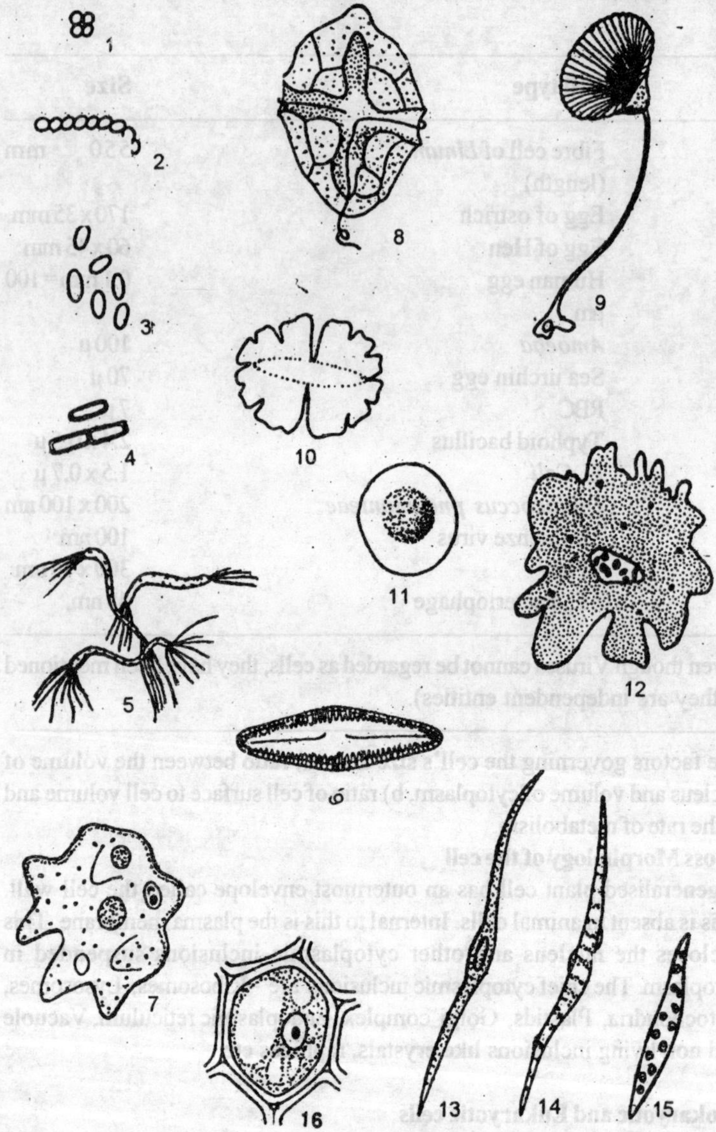

Fig. 3.1 The Cell

1. Tetracoccus, **2.** Streptococcus, **3.** Micrococcus, **4.** Bacillus, **5.** Spirillum, **6.** Navicula (Diatom), **7.** Amoeba, **8.** Dinoflagellate, **9.** Acetabularia, **10.** Desmid, **11.** Human Erythrocyte, **12.** Melonocyte, **13.** Smooth muscle cell, **14.**triatedmuscle cell, **15.** Trachied from a vascular plant, **16.** A parenchyma cell

does not form definite chromosomes in prokaryotic cells. The DNA is circular and lies in a tangled mass. In eukaryotes, the chromenema forms definite number of chromosomes and is associated with proteins.

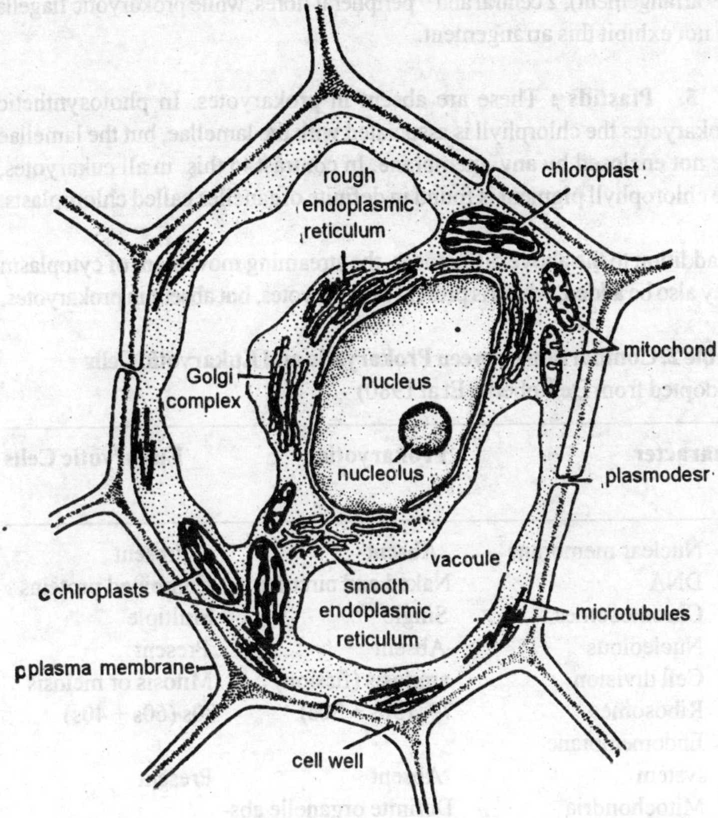

Fig. 3.2 The Cell

Structure of a typical Plant cell (Electron Microscopic view)

2. Cytoplasmic organelles : Mitochondria, Endoplasmic reticulum, Golgi complex, lysosomes and centrioles are absent in prokaryotes. The enzymatic functions associated with mitochondria are carried out by the infolding of the cell membrane at several points. Eukaryotes have all the organells mentioned above.

3. Cell wall : Chemically the cell wall of prokaryotes is composed of amino sugars and muramic acid. These are absent in the cell wall of eukaryotes (plants).

4. **Flagella;** The flagella of eukaryotes have eleven fibres with (2 + 9 arrangement), 2 central and 9 peripheral fibres, while prokaryotic flagella do not exhibit this arrangement.

5. **Plastids :** These are absent in prokaryotes. In photosynthetic prokaryotes the chlorphyll is associated with the lamellae, but the lamellae are not enclosed by any membrane. In contrast to this, in all eukaryotes, the chlorophyll pigment is found in definite organelles called chloroplasts.

In addition to the above differences, the streaming movement of cytoplasm may also be added, which is present in eukaryotes, but absent in prokaryotes.

Table 2. Comparison between Prokaryotic and Eukaryotic Cells
(Adopted from De Robertis Et al 1980)

Character	Prokaryotic	Eukaryotic Cells
1. Nuclear membrane	Absent	Present
2. DNA	Naked and circular	Combined proteins
3. Chromosomes	Single	Multiple
4. Nucleolous	Absent	Present
5. Cell division	amitosis (fission)	Mitosis or meiosis
6. Ribosomes	70s (50s + 30s)	80s (60s + 40s)
7. Endomembrane system	Absent	Present
8. Mitochondria	Definite organelle absent,enzymes of respiration andphotosynthes- is stored in theolds of plasma membrane	
9. Choloroplast	Absent	Present
10. Cell wall	Non cellulosic(present in plants only)	Cellulosic
11. Exocytosis and endocytosis	Absent	Present
12. Flagella	No definite arrangement of fibrills	9 + 2 fibrillar arrangement

PROKARYOTIC CELL

The prokaryotic cells are generally smaller and vary in size in different members. In mycoplasma it is about 0.12 µm while in *Oscillatoira,* a filamentous blue green the size is 40 x 5µ. A great majority of them, however are about (1.0µm) in size. Among the bacteria the smallest are to be found among cocci (0.1 µ) while the largest are the spirilla (60 x 6~m). The various components of the prokaryotic cell are described below.

Cell wall : Cell wall is absent in some prokaryotic (mycoplasma, L forms of bacteria) members while it forms the outermost boundary in a large majority of them. While the cell wall is non- living and does not have the property of selective permeability it can act as a seive as it is porous.

In some prokaryotes the space between the cell membrane and cell wall is filled with some enzymes and metabolites consituting the periplasm.

Chemically the cell wall is made up of strong fibres of hetropolymers called *mucopeptides of peptidogeycans.* They are also refered to as *glycopeptide, muropeptide, murein* etc.

The peptidoglycans consist of alternating units made up of N acetyl glucosamine and N acetyl muramic acid with 1-4 linkages. This type of a linkage is common to all prokaryotes. Attached to the muramic acid is a short peptide chain consisting of alanine, glutamic acid and either lysine or diaminopimelic acid (DAP) and muramic acid. DAP and some of D-amino acids are found only in prokaryotic cell walls.

Cross linkage of these polymers assures structural rigidity with the type and extent of linkage being species specific. For examples in some bacteria a pentaglycinc links the tetrapeptide side chains extending from the muramic acid units while in others the terminal D- alanine of one tetrapeptide may be covalently linked to the DAP of an adjacent tetrapeptide. The peptidoglycans of gram negative bacteria have more cross linkage than those of gram positive bacteria.

In addition to peptiodglycans the wall has many other chemicals. The wall of gram negative bacteria shows as many as five layers under the electron microscope. These are–the periplasm space, Peptidoglycan layer, an intermediate zone, the outer membrane and lipopolysaccharides.

Cell membrane : This is the bounding layer of cytoplasm and is about 6-8nm thick. Unlike in the case of eukaryotic cells it carries out multiple functions. Some of these are –

(i) Presence of enzymes of biosynthetic pathways leading to the synthesis of wall components
(ii) Occurrence of *pennease* enzymes responsible for the transport of materials across the membrane
(iii) Membranes of aerobic bacteria have the electron transport chain and oxidative phosohorylation
(iv) The purple bacteria have their photosynthetic apparatus in the membrane
(v) The membrane has attachment sites for chromosomal DNA (also plasmid) during DNA partitioning
(vi) Regulatory mechanism for chemotaxis is located in the membrane.

Structurally the membrane is triple layered consisting of two electron dense layers surrounding an electrontranslucent layer. In otherwords, it is similar to a unit membrane. The architecture and chemical composition is similar to the fluid mosaic model proposed by Singer and Nicolson (1972) (For details see later in the chapter).

Cytoplasm : Cytoplasm does not appear to have an organized structure when observed under an electron microscope. It may be regarded as a concentrated solution containing a host of metabolites, enzymes, amino acids, inorganic ions, sugars, vitamins etc. Cytoplasm plays an important role in intermediary metabolism. There is no cytoskelcton.

Flagella (see bacteria)
Capsule (see bacteria)
Pili (see bacteria)

Surface layers : These are present in all gram negative and gram positive bacteria. Three main types of S layers have been reported in prokaryotic cells (bacteria). The S layers are composed of single homogenous peptides with occasional carbohydrates as minor components. These layers are believed to act as macromolecular sieves preventing the entry of toxic molecules and also to protect the peptidoglycans of the wall from the action of lytic enzymes.

Nucleoid : This represents the genetic material of prokaryotic cells. It is similar in all prokaryotes. (for details see - bacteria) The molecular structure of DNA is the same as in eukaryotic cells.

Inclusion bodies : These are storage granules and other particles distributed in the cytoplasm. There are two kinds of inclusion bodies–membrane enclosed and membraneless (Shiviey 1974).

Membrane enclosed bodies : These are enclosed by a non-unit type of membrane, 2-4 nm, thick and made up of proteins. The inclusion bodies encountered in the prokaryotic cells are chlorosomes, carboxysomes, magnetosomes, gas vacuoles, polyhydroxybutyrate (PHB) granules, sulphur granules and glycogen granules.

PHB granules occur in *cyanobacteria*, sulphur granules are found in *Thiorhodaccae* and glycogen granules in *Clostridium*.

Chlorosomes have been demonstrated in photosynthetic bacteria belonging to chlorlobiaceae. (Cruden and Stanier, 1970). These are different from chromatophores in being non-membranous.

Carboxysomes are seen in cyanobacteria (Murphy et al, 1974).

Magnetosomes are particles which help the bacterium in orienting itself in relation to the magnetic behaviour of the surroundings. Reported to occur in *Aquaspirillum magnetotacticum* (Balkwill et al, 1980), the magnetosomes allow the bacteria to punise the most efficient aerotactic behaviour.

Non membranous Inclusion bodies: Phycobilisomes, cyanophycin granules (found in cyanobacteria), polyglucoside granules and several crystalline substances belong to this group.
In addition to the above, there are R - bodies (retractile stuctures) demonstrated in bacteria by Lalucat and Mayer (1978). These are supposed to be similar to kappa particles of *Paramecium*.

Ribosomes (see bacteria)

Mesosomes (see bacteria)

Chromatophores : These are present in the photosynthetic bacteria belonging to the families– Rhodospirillaceae, Chromatiaceae and Cyanophyceae. Chromatophores vary in form and have vesicles, tubes, bundles, stacks or thylakoids etc.

EUKARYOTIC CELL

Cell Wall

The cell wall is the characteristic feature of the plant cells. It is made up of non living matter and is believed to be secreted by the plasma membrane. The structure and composition of the cell wall not only varies in different plants, but also in different types of cells of the same plant. Often, there might be a variety in the cell wall composition at different stages of the growth of the cell. The cell wall is very rigid and maintains the shape of the cell acting as a boundary. A unique feature of the cell wall is its permeability which is due to the presence of small pores.

Gross structure

Three layers are distinguishable in the cell wall. These are the *primary cell wall, middle lamella* and the *secondary cell wall.* The primary cell walls of adjacent cells are separated by middle lamella. The secondary wall is deposited after a certain period of development.

The primary wall is secreted by the plasma membrane. In some instances like the leaves, fleshy stems, roots etc., the cells have only the primary wall and the middle lamella. The primary wall is formed during the early stages of growth and is about 1 to 3nm in thickness. Chemically the primary wall is composed of cellulose, hemicellulose and pectic compounds. The young primary wall is plastic and is prone to changes. Often it may be lignified.

The middle lamella is the intercellular substance binding together the primary and the secondary walls. The middle lamella is formed during cell division and is amorphous and colloidal in nature. Chemically the middle lamella is composed of pectin, cellulose, calcium and magnesium pectate.

The secondary wall is deposited on the primary wall after cell maturation. It is very rigid and does not alter its shpae. The thickness varies from 5 – 10 nm. The secondary wall usually has three layers viz., *outer layer, middle layer* and *inner layer.*

Chemically the secondary wall is composed of cellulose, non cellulosic polysaccharides and hemicelluloses.

Ocassionally a *tertiary cell wall* may also be found as in the tracheids of some gymnosperms. This is a very thin layer and is composed of *xylan,* instead of cellulose. The secondary wall acquires different types of thickenings. (For details see *College Botany* Vol II - S. Sundara Rajan)

Ultra structure of the cell wall

The cell wall is mainly composed of cellulose. Cellulose is a polymer of glucose molecules. Approximately each molecule of cellulose consists of about 3000 glucose residues joined together by 1-4 oxygen bridges. These cellulose chains are oriented along the axis of the cell. About 100 of these

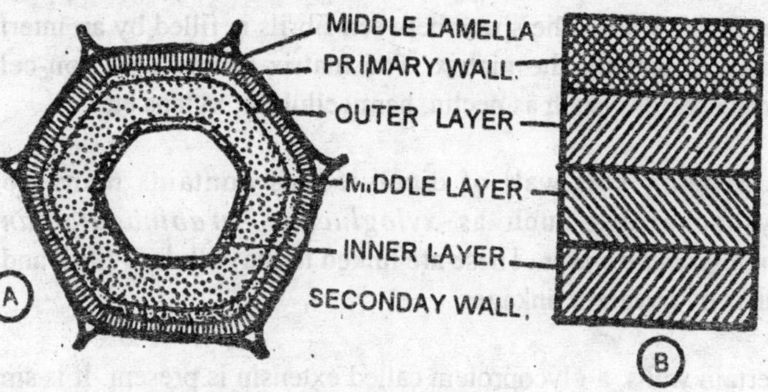

Fig. 3.3 The Cell

Arrangement of Wall layers. A. Cross section, B. Arrangement of fibrils cellulosechains bundle together to form *micellels* or *elementary fibrils.*

Each micelle is about 100Å in diameter. Many hundreds of these micelles of these stack together to form *microfibrils*. Each microfibril is 200Å to 300Å in diameter. The micorfibril is visible only under an electron microscope. About 2000 of these microfibrils join together to constitute the *macrofibril* which is about 0.5μ in width. The macrofibrils are visible under the light microscope.

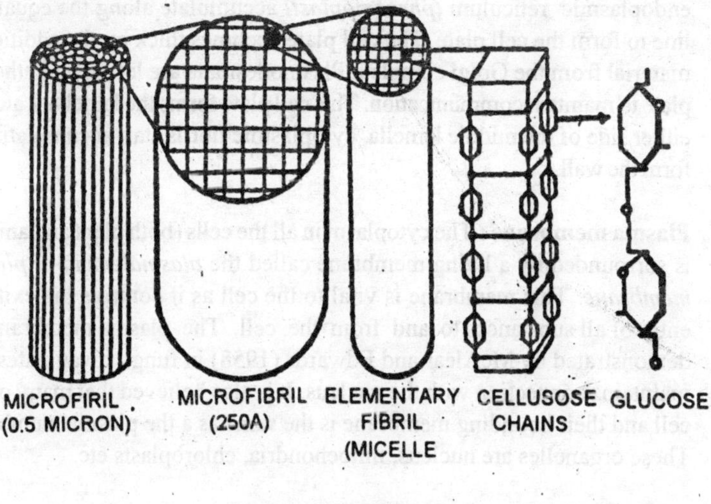

Fig. 3.4 The Cell

Structure elements of Cellulose (An important component of plant cell walls)

In secondary walls, the microfibrils are more orderly and compact than in the primary walls. The space between fibrils is filled by an interfibrillar substance, usually the matrix. The matrix is filled by non-cellulosic polysaccharides–such as pectin, hemicellulose, xylans etc.

The primary cell wall of dicotyledons contains many complex polysaccharides such as *xyloglucans, arabinogalactans* and *rhamnogalacturonans.* These are linked to the cellulose fibril and to one another by covalent linkages.

In certain walls, a glycoprotein called extensin is present. It is similar to collagen in being rich in 4 - hydroxyproline.

Extensin has a long protein chain, to which are linked the hydroxy groups of residues of *arabinose* and *galactose.*

Plasmodesmata : Protoplasts of adjacent cells are connected to one another by means of thin cytoplasmic threads called plasmodesmata. These pass through the small pores (pit fields) present in the cell wall. Tubules which are in continuity in the endoplasmic reticulum are seen within plasmodesmata, thus maintaining an effective intercellular communication.

Origin of cell wall : At the completion of *Karyokinesis,* small vesicles of endoplasmic reticulum *(phragmoplast)* accumulate along the equatorial line to form the cell plate. The cell plate becomes thick by the addition of material from the Golgi complex. Plasmodesmata are left open in the cell plate to maintain communication. This cell plate forms the lamella. Later, on either side of the middle lamella, cytoplasmic fibrils start accumulating to form the wall.

Plasma membrane : The cytoplasm in all the cells (both plant and animal) is surrounded by a living membrane called the *plasmalemma* or *plasma membrane.* This membrane is vital to the cell as it controls the exit and entry of all substances to and from the cell. The plasma membrane, as demonstrated by McAlear and Edwards (1958) in fungi, invaginates and maintains connection with the nucleus. It is also believed that many of the cell and their bounding membrane is the same as a the plasma membrane These organelles are nucleus, mitochondria, chloroplasts etc.

The presence of the cell membrane is proved by the following evidences.

1. On injury to a cell, when protoplasm flows out, it remains as a discreet unit.
2. Observations under electron microscope have revealed the presence of a boundary membrane over the surface of cytoplasm.

Size of the membrane : It is about 75Å thick and is composed of three concentric layers. The middle layer is about 35Å in thickness, while the outer layers on either side are 20A thick.

Chemical composition : Membranes from different sources show variations in chemical composition. Essentially however, all membranes are lipoproteinanceous. The components are held together in a thin sheet by non convalent bonds. The ratio of protein to lipids varies depending on the type (Golgi membrane, plasma membrane etc) of membrane, type of organism (eukaryote, prokaryote, etc.) and the type of cell epidermis, phoelm etc.). In addition to these two, certain membranes also have carbohydrates.

Membranes have three different categories of proteins. These are – *structural proteins, functional proteins* (enzymes) and *carrier proteins. Structural* proteins maintain the frame work of the membranes. The average molecular weight of structural proteins found in membranes is $3x10^4$. Enzymes catalyzing various physiological activities are found in the membrane of mitochondria, endoplasmic reticulum etc. *Carrier proteins* are known to form a complex with the substances and help them to pass through the membrane.

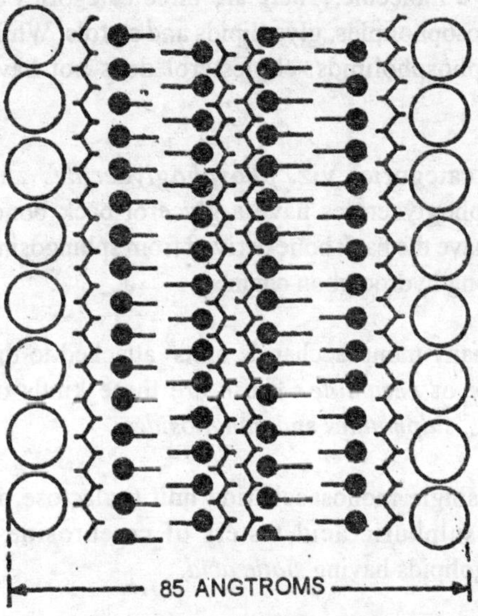

Fig. 3.5 The Cell
Diagrammatic representation of a cellmembrane having two unit membranes

The protein composition of different membranes vary (as has been pointed out already). For instance, plasma membranes have 50% while mitochondrial membranes have 75% proteins.

The proteins present in plasma membranes are categorized into two viz. *integral proteins* and *peripheral proteins*. While integral proteins form the back bone of the membrane, peripheral proteins have a weak association maintained by weak electrostatic interaction. The integral proteins are hydrophobic in nature as they have to penetrate the hydrophobic lipids to form the basic structure of the membrane. These proteins do not easily disassociate from the membrane and it requires harsh chemical procedures to accomplish it. The integral proteins have a hydrophilic portion also, which protrudes out of the lipoid bilayer. Peripheral proteins are associated either with the hydrophilic head of lipids or hydrophilic part of the integral proteins.

The lipids of membranes are of several types, but all of them share a common character in being *amphipatic* i.e., having both *hydrophilic* and *hydrophobic* portions within a molecule. There are three categories of membrane lipids–they are phosopholipids, glycolipids and sterols. While most membrane lipids are phospholipids, cholesterol does not have phosphates.

Phospholipids are of two categories viz. *phosphoglycerides* and *sphingophospholipids*. Phosphoglycerides have a glycerol back bone, while sphingophospholipids have the back bone derived from sphingosine an aminoalchobol having a long hydrocarbon chain.

Glycolipids contain one or many monosaccharide units attached to the lipids based on *sphingenine* or *ceramide*. These are three kinds of glycolipids viz., *Cerebrosides, sulphatides* and *gangliosides.*

Cerebrosides contain a single monosaccharide unit (galactose or glucose). Sulphatides are sulphuric acid testers of cerebrosides. Gangliosides are glycosphingolipids having *sialic acid.*

Sterols are steroid alcohols and are of three types – cholesterols (present in animal tissues), phytosterols (present in plants) and ergosterols (present in eukaryotic microorganisms).

Among the glycerophospholipids, *lecithin* and *cephalin* are the most abundant in higher plants and animals.

Membrane functions

The following are the major functions of membranes.

1. **Compartmentalization:** Membranes constitute the living boundary for cytoplasm and cell contents. As unbroken sheets, they divide the living matter into self sustaining units to effectively coordinate and regulate the activities.

2. **Regulation of movement:** In a biologically active cell many susbstances enter and many others go out of the cell. However, if the entry and exit is unregulated, it is deleterious to the cell. Hence, plasma membrane regulates and decides which substances are to be allowed and which should not. This unique property of the cell membrane is called selective permeability or differential permeability is responsible for most of the physiological activities of the cell.

3. **Communication:** All the membranes have certain special substances (on their surface) called receptor molecules which mediate and control the message from one to another. The receptor molecules vary in different membranes. These molecules combine with substances (ligands) of different types and generate signals (in the form of cyclic AMP) directed to the interior of the cell. The best studied ligands are the hormones and neurotransmitters.

4. **Interaction between cells:** It is the responsibility of the plasma membrane (as the living boundary between cells) to effectively coordinate actions between cells since an organism consists of differnt types of cells, and many a time the actions have to be coordinated to achieve harmony.

5. **Energy transformations :** Membranes are intimately associated with the interconversion of energy. For instance, in green plants the light energy has to be converted into chemical energy. Similarly the chemical energy stored in carbohydrates has to be converted into energy currency i.e. ATP molecules. The machinery for energy capture and conversion is hidden in the membranes of mictochondria and chloroplasts.

6. **Locus for biochemical functions :** Membranes provide the structural framework for the location of various enzymes. The scaffolding of the membrane interconnecting the various organelles helps in the coordination of activities. Such an effective coordination of activities is best illustrated in the mitochondrial membrane where the various components of the

electron transport chain work together as they are located in an orderly fashion.

7. Other activities : Membranes also function in various other capacities such as providing mechanical strength, involving in cell movement secretion etc., and also acting as electrical insulators.

Structure of the membrane : The molecular structure of the membrane has been interpreted in various ways by different cell biologists in order of their known functions. Various models have been proposed for the molecular structure of the membrane. These fall into two basic categories, viz., bilayer models and micellar or subunit models. The bilayer models are believed to have protein and lipid arranged in the form of layers, while the subunit models assume the presence of a number of independent subunits. We shall discuss a few of these models. The chart below summarises the list of various models.

MEMBRANE MODELS

- Bilayer models
  - Lipid membrane Model
  - Sandwich Models
    - (a) Danielli Davson model
    - (b) Unit membrane model
    - (c) Greater membrane odel
    - (d) Kavanau's model
- Micellar models
  - Models in which protein penetrate into the lipid layer
    - (a) Benson's model
    - (b) Lenard and Singer' model
    - (c) Mosaic membrane model
    - (d) The fluid mosaic model

BILAYER MODELS

Lipid membrane model : Overtoil (1895), discovered that fat soluble subsatances easily pass through the membrane indicating their affinity to the lipids. Hober (1910) and Fricke (1925) found out that the low electrical conductivity of the Iving cell is due to the presence of lipids. The experiments of Mudd and Mudd (1931) by placing an oil droplet which adheres to the cell surface provided the presence of lipids in the cell membrane. It was Gorter and Grendel (1925) who suggested a structural model based on lipids. They studied the erythrocyte (RBC) membrane and concluded that it is composed of two layers of liquid molecules whose polar groups are situated on the outside of the membrane.

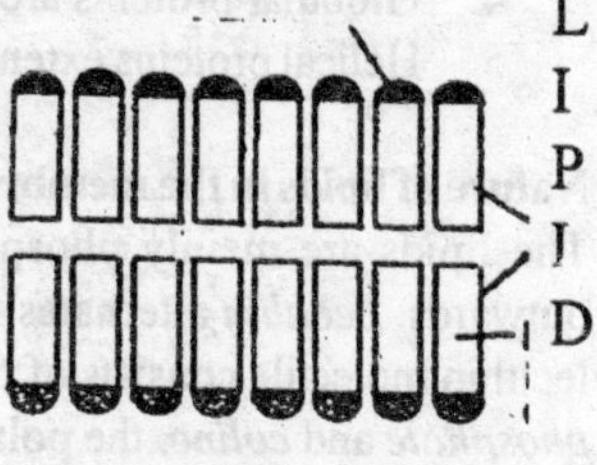

Fig. 3.6 The Cell
The lipid membrane model of Forter and Grendel

Sandwich models : In these models the lipid layers are supposed to besandwiched between the two protein on either surface. The polar groups of the lipids adhere to the proteins on either side. The following are some of the models based on sandwhich principle

(a) Danielli Davson model : The study of surface tension of cells by Daniell and Harvey (1935) and Harvey and Cole (1931) indicated that there must be more to the structure of the membrane than merely being made up of lipids. In 1935, Hugh Davson and James Denielli proposed that the plasma membrane is a lipid layer sandwiched between layers of globular proteins. In a way the Danielli model was the first structural interpretation to account for the physiological properties of the membrane.

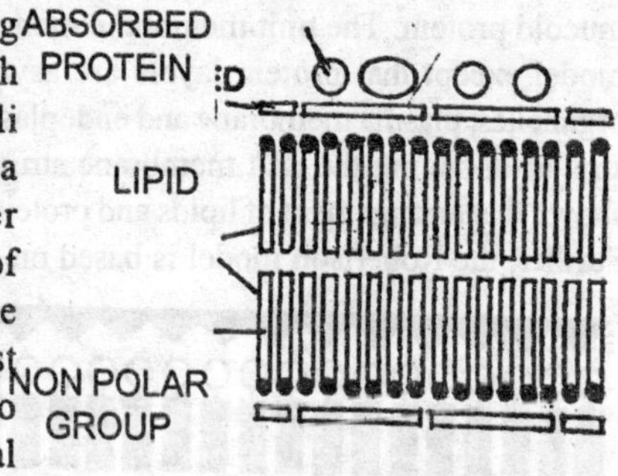

Fig.3.7 The Cell
The Danielli-Davson model

Over the years various modifications have been proposed for the model. Danielli (1938) opined, that there could be two kinds of globular proteins on the surface and diagonally arranged proteins in contact with the lipid layer. In order to account for the permeability, certain small pores were visualised in the lipid layers with an inner lining of protein molecules providing an inner hydrophilic surface.

The following are some other variations in the protein lipid arrangement.

1. Proteins are in the form of folded chain.
2. Proteins are in a coiled form on both sides of the lipid layer.

3. Proteins are asymmetrically arranged i.e., one side has globular proteins and the other side a folded β chain.
4. Globular proteins are found on either side.
5. Helical proteins extend into the pores between the lipid molecules.

Nature of lipids in the membrane

The lipids are mainly phospholipids with their polar groups pointing outwards. *Lecithin* alternates with a steroid molecule (*cholesterol*). Each lecithin molecule consists of two glycerol chains with the part containing *phosphate* and *coline,* the polar head is bent like a walking stick.

(b) The unit membrane model : The specimens of membrane fixed in osmium tetroxide for the purpose of electron microscopic observation reveal the occurrence of *two lines,* presumably indicating their double membrane nature. This observation originally made in 1950's by Humberto Fernandez Moran and David Robertson led to the unit membrane model. Robertson used a new fixative, potassium permanganate. He discovered that each of the dense line possessed a sub structure having a trilaminar appearence. Each membrane is composed of two thin electron dense lines of 20Å, with the intervening space of about 35Å. The inner and outer layers are non-identical. The outer is a mucoid protein, while the inner surface is a non-mucoid protein. The unit membrane model is similar to Danielli Davson model except that protein layers are asymmetrical. Membranes of cell organelles, plasma membrane and endoplasmic reticulum are thought to be similar in having the unit membrane structure. But current studies have shown that arrangement of lipids and proteins vary in different membranes. Further, the Robertson model is based on the study of myelin which is a

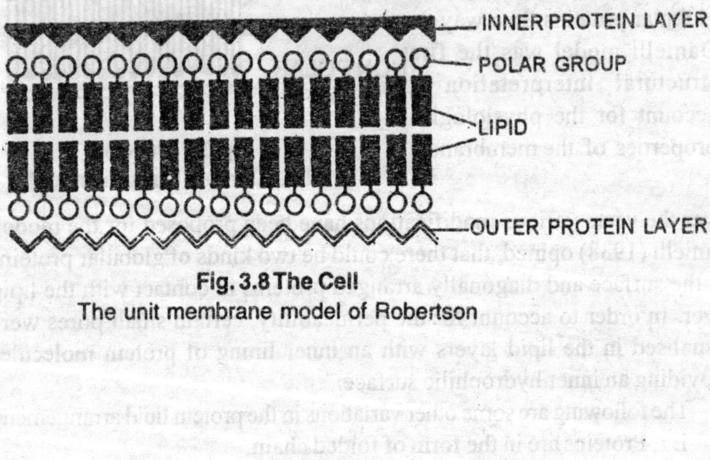

Fig. 3.8 The Cell
The unit membrane model of Robertson

(c) The Greater membrane model : This is similar to the trilaminar sandwich model except that the outer and inner surface proteins are different. The inner surface has an unconjugated structural protein adhering to polar heads of lipids. The outer surface also has structural proteins, on which is superimposed a *glycoprotein.* To this *glycoprotein* are attached side chains of oligosaccharides.

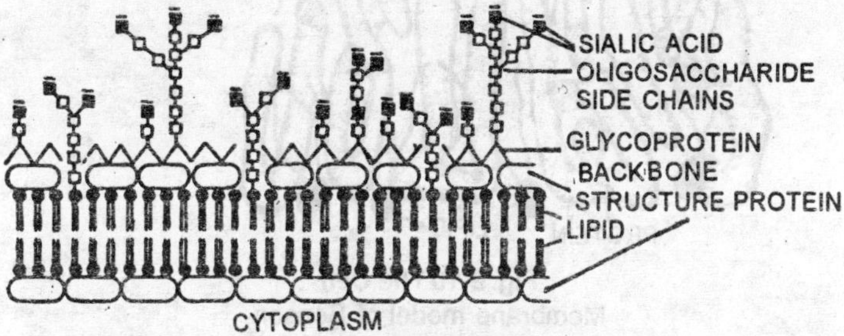

Fig. 3.9 The Cell
The greater membrane model

(d) Kavanau's lipid pillar model : According to Kavanau J.L. (1965), the lipid layer exists in two forms. Kavanau's model is essentially similar to Danielli Davson model. The two types of lipid structures are *pillars* and *flattened discs.* Pores are present between pillars to faciliate the passage of ions. Membranes with pillar lipids are thicker than those with discs. Internally the pillars have non-polar tails, while the surface is made up of polar phospholipids. Proteins are present on either side of the lipids.

Models based on protein penetration of lipids

(a) Lenard and Singer's model : This is based on the protein arrangement in the membrane. About 25 to 30% of proteins are helical while the rest are irregular coils.

(b) Benson's model : This model was proposed by Benson (1966), based on the study of membranes of chloroplasts. In this, it was postulated that lipid tails are bound to the corresponding hydrophobic points within the coils of proteins. The polar heads of the lipids lie on the surface.

(c) Mosaic membrane concept : According to this model proposed by Baum (1967), the membrane consists of units, each about 80Å in diameter. The wedge shaped protein is enclosed on either side and at edges by the lipids. The polar heads protrude out combining with the hydrophobic areas of proteins.

(d) The Fluid mosaic model : The molecular model of the plasma membrane proposed by Singer and Nicolson (1972), emphasises on the dynamism of membrane organisation and visualises it to be a semifluid like structure with a mosaic arrangement of lipids and proteins. The fluid mosaic model, though essentially based on bilayer models, rejects the idea of continuous bilayer.

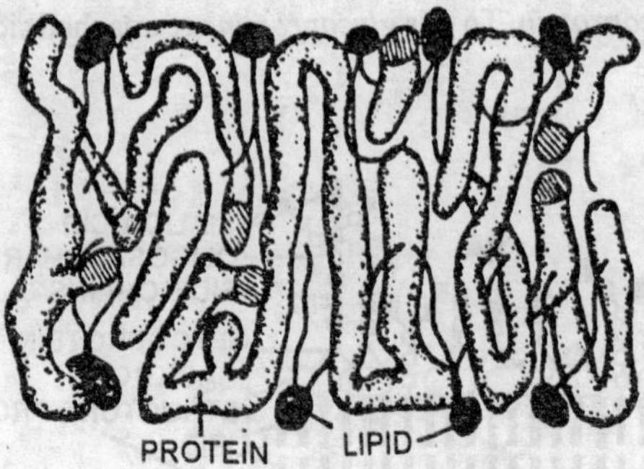

Fig. 3.10 The Cell
Membrane model of Benson

Instead of forming continuous external layer adhering to the hydrophilic heads of the phospholipids, as in the Danielli model; the proteins constitute discontinuous particles often penetrating deeply into the lipid layer. In a way, the protein units could be compared to floating icebergs in a sea of phospholipids. The countinuity of the bilayer is not maintained because of complete or incomplete interruption by proteins. While the Danielli model assumes a hydrophilic bonding between lipids and proteins, the Singer Nicolson model regards it to be hydrophobic.

The membrane proteins in the fluid mosaic model are of two types distinguishable by their intensity of lipid association. These are the *integral* or *intrinsic proteins* and the *extrinsic* or *peripheral proteins*. The integral proteins penetrate deep into the lipid layer and are the hydrophobic as they have to coexist in the hydrophobic environment of fatty acids. The integral proteins are *amphipatic*. While their tips are hydrophilic; the central portion is hydrophobic. These proteins are difficult to extract from the membranes.

The peripheral proteins which are weakly associated with the membrane can be easily extracted. They are hydrophilic and never penetrate the interiors of the lipid bilayer. They form weak bonds either with the polar heads of lipids or with the hydrophilic surfaces of integral proteins. One of the best evidences for a broken surface layer of proteins comes with the use of phospholipases. These phosphate degrading enzymes very easily percolate into the membrane, which would have been impossible, if there were to be an unbroken surface envelope of proteins.

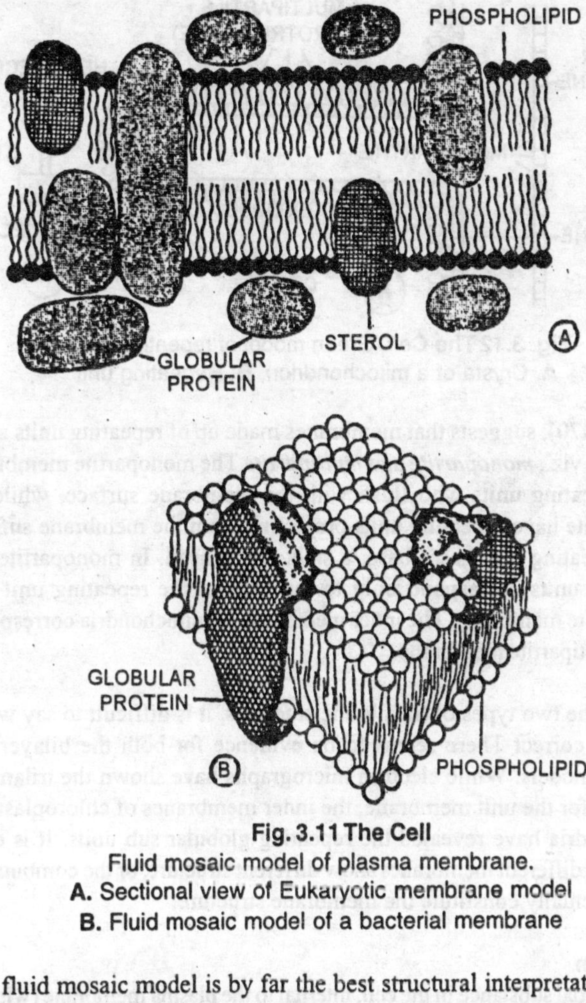

Fig. 3.11 The Cell
Fluid mosaic model of plasma membrane.
A. Sectional view of Eukaryotic membrane model
B. Fluid mosaic model of a bacterial membrane

The fluid mosaic model is by far the best structural interpretation of the membrane which satisfactorily explains all its (membrane's) known functions.

MICELLAR MODELS

According to Fernandez Moran (1963), the membrane is corpuscular in nature consisting of many repeating units. These units called *globular units* or elementary particles range in diameter form 40 - 70Å and form closely packed repeating units. The membrane structure may be globular or unit membrane in different organelles or under different metabolic conditions. According to some cell biologists, the corpuscular nature is an artifact.

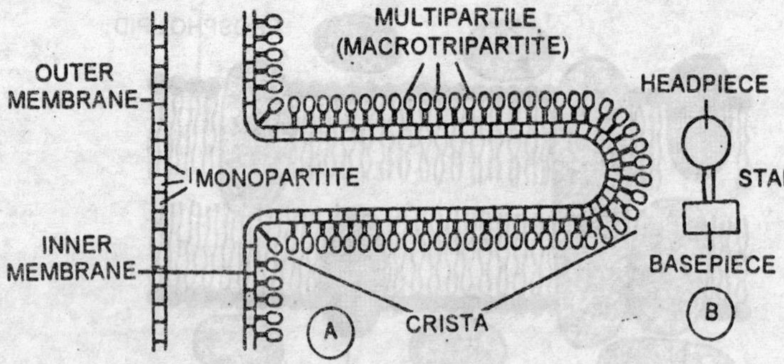

Fig 3.12 The Cell. Green model of repeating units
A. Crysta of a mitochondrion, **B.** Repeating unit

Green (1970), suggests that membranes made up of repeating units are of two types viz., *monopartite* and *multipartite* The monopartite membranes have repeating units who flush with the membrane surface, while the multipartite have repeating units projecting from the membrane sufrace. Each repeating unit has a base, a stalk and a head. In monopartite, the repeating units correspond only to the base of the repeating unit of a multipartite membrane. The inner membrane of mitochondria corresponds to the multipartite membrane.

Of the two types of membrane stuctures, it is difficult to say which model is correct There seems to be evidence for both the bilayer and micellar models. While electron micrographs have shown the trilaminar structure for the unit membrane, the inner membranes of chloroplast and mitochondria have revealed the repeating globular sub units. It is quite likely that different membranes show different structure, or the combination of both actually constitute the membrane structure.

Cytoplasm

The colloidal substance in the cell, internal to the plasma membrane (without the inclusions), is called cytoplasm. The cytoplasm is usually granular and has many substances either dissolved or suspended in it The suspended particles may be in the solid or semisolid condition. There are several theories regarding the structure of the cytoplasm. Some of these are–

Reticular theory : Cytoplasm is made up of a network of fibrils or particles.

Alveolar theory : According to Butschill (1892), cytoplasm consists of a number of droplets or alveoli. It gives an appearence of oil droplets floating on water.

Granular theory : According to Altmann (1893), cytoplasm is composed of a number of minute granules suspended in a matrix.

Fibrillar theory : This theory was proposed by Fisher (1894), and Flemming (1897). Electron microscopic studies have proved the hypothesis of Fisher. According to this view, cytoplasm is made up of a number of fibrils.

Properties of cytoplasm : Cytoplasm has a number of physical and biological properties.

Physical properties

The colloidal nature of cytoplasm is responsible for the various physical properties A colloidal system may be defined as a system of liquid medium containing suspended particles varying in diameter from 1/1,000,000 to 1/10,000 nm. Some of the physical properties are as follows.

Phase reversal : Cytoplasm may be differentiated into a *sol state* and *gel state.* Sol state is the liquid state and the gel state is the semi-solid state. These two phases are interchangeable. This capacity is known as phase reversal.

Elasticity : Depending on the circumstances, the cytoplasm can extend or contract subject to a certain limit. This is known as elasticity.

Cohesiveness : The suspended particles in cytoplasm have mutual attraction, thus exhibiting cohesiveness.

Contractility : It is the capacity of peripheral cytoplasm to absorb or remove water from cell to the exterior. This is manifested very well by guard cells.

Viscosity : The suspended particles of cytoplasm are responsible for its viscous nature.

Brownian movement : The suspended particles of cytoplasm are, in a state of to and fro movement, called Brownian movement.

In addition to this, cytoplasm also exhibits movements (seen in the plasmodium of slime molds) and 'cyclosis' or streaming movements (seen in the leaf cells of *Elodea).*

Biological properties

Growth, respiration, excretion, nutrition, reproduction, metabolism, irritability etc., are some of the important biological activities of cytoplasm.

Chemical composition of cytoplasm : 90% of cytoplasm is water. The remaining 10% is constituted by organic and inorganic compounds.

About 40 or more inorganic constituents are found in cytoplasm. These are found in the form of ions (either cations or anions). Some of the common inorganic constituents are – Oxygen (62%), Hydrogen (10%), Carbon (18%) and Nitrogen (3%). The less common inorganic contituents are Calcium (0.25%), Phosphorus (1.2%), Sodium (0.1%), Magnesium (0.7%), Iodine (015%) and Chlorine (0.15%).

Among the various organic substances proteins, carbohydrates, lipids, hormones, vitamins and nucleic acids are most important.

Cytoplasmic Inclusions : Cytoplasm has a number of living and non-living inclusions. The living inclusions are – Nucleus, mitochondria,. plastids, endoplasmic reticulum, golgi complex, lysosome, centrioles, microbodies, etc.

Nucleus

This is the most important organelle of the cell, which regulates all its activities. It was discovered by Robert Brown in 1831.

Position : Usually it is present at the centre of the cell, but, in some instances, as in mature parenchyma cells, it is at the nucleus is peripheral. In *Acetabularia,* the nucleus occupies various positions, though mainly present at the basal region.

Shape, size and number : The shape is variable depending on the cell type. Though the nucleus is generally spherical, or ellipsoidal flattened ones also may be seen in certain cells. The surface is generally smooth but, as in leucocytes it may have small infoldings.

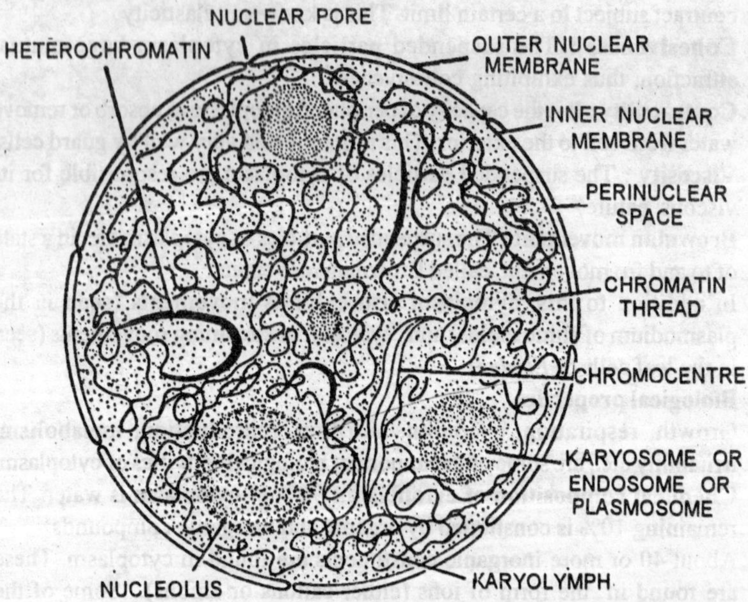

Fig. 3.13 The Cell
Structure of the nucleus

The size of the nucleus is also variable depending on the cell; the size of the nucleus being directly proprotional to that of cytoplasm. The following formula or Heywig may be used to determine the size of the nucleus in a particular cell.

$$NP = \frac{Vn}{Vc\text{-}Vn}$$

When *NP* = nucleoprotein index, *Vn* is the volume of the nucleus and V*c* is the volume of the cell.

Most of the cells are uninucleate, while some *(Paramecium)* may be binucleate or multinucleate *(Cladophora)*. In certain fungi like *Mucor*, where there are no septa dividing the plant body into cells, numerous nuclei occur indicating a coenocytic condition.

Ultrastructure of the nucleus : Ultrastructurally the nucleus has the following parts – 1. Nuclear membrane, 2. Karyolymph or Nuclear sap, 3. Chromonemata, 4. Nucleolus and 5. Endosomes.

Nuclear membrane : In all eukaryotic organisms, a limiting membrane envelops the nucleus. This membrane helps in effective communication between nucleus and cytoplasm. The elements of endoplasmic reticulum contribute to the nuclear envelop during cell division.

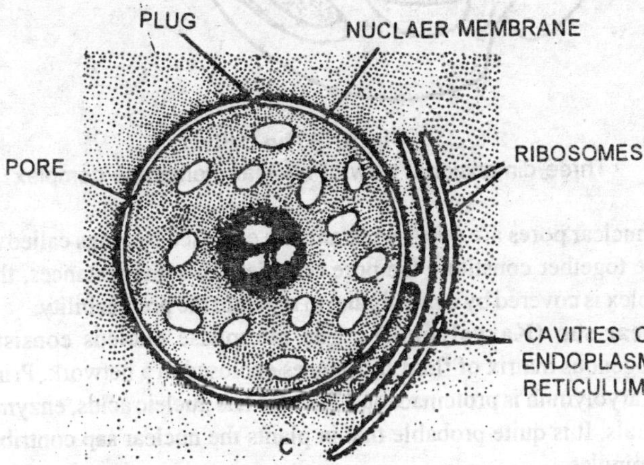

Fig. 3.14 The Cell

Diagrammatic representation of nuclear membrane and pore

Electron microscopic studies have revealed, the nuclear membrane to be a double membrane with a number of pores. According to Watson (1959), these pores called *nuclear pores* are octogonal in shape. The outer nuclear membrane is called endokaryotheca. Each membrane is about 75 - 90Å thick and they enclose a space between them about 100 - 150Å (De Robertis et al 1970). This space is called the *perinuclear space* or *cisterna*. The outer membrane is rough due to the presence of RNA particles. Internally the nuclear membrane is followed by a supporting membrane called *fibrous lamina* (300Å thick).

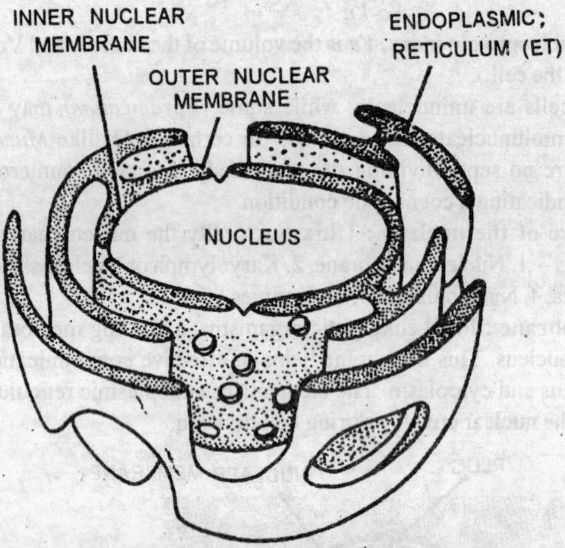

Fig. 3.15 The Cell
Three dimensional view of nuclear membrane complex

The nuclear pores are 600Å wide and are enclosed by rings called *annuli.* These together constitute the pore complex. In some instances, the pore complex is covered by a membrane to regulate the premeability.

Nuclear sap (Karyolymph) : The interphase nucleus consists of a homogenous matrix of fluid substances enclosed in a network. Primarily, the Karyolymnh is proteinaceous, but also has nucleic acids, enzymes and minerals. It is quite probable that in plants the nuclear sap contributes to the spindles.

Chromonemata : Enclosed in the Karyolymph and visible in the interphase nucleus, are found a number of fibrillar structures constituting a network called *chromonemata* or *chromatin fibrils.* These, stain positively with basic fuchsin stain. Some coarse granules are deposited on the chromatin network. These are called *chromocentres* and constitute the points of condensation of chromosomes. During cell division, the chromatin network breaks up into a specific number of chromosomes. Two regions can be identified in the chromatin material. These are *heterochromatic* region and *euchromatic* region.

The heterochromatic region stains darkly and shows numerous bead like structures called *chromomeres.* The heterochromatic region has less DNA. This region is believed to be genetically and metabolically inert. The light staining region of the chromatin is called the euchromatic region. This region contains more of DNA and is supposed to be genetically active.

Nucleolus : This was first discovered by Fontana (1874). A spheroidal body, situated either in the central or pheripheral position, the nucleolus is supposed to regulate the synthetic activity of the nucleus. Usually two or more chromosomes are associated with the nucleolus (this can be seen during late prophase), and these are called *nucleolar organisers* as they play a role in the reappearence of the nucleolus after cell division. The number of nucleoli or nucleus varies from one to two or three.

Chemically the nucleolus is rich in RNA. According to Palade (1963), the nucleoli of liver cells contain 3–5% RNA, while pea embryos have 10–20% of nuclear RNA. Orthophosphates helping in RNA synthesis have been reported in nucleolus. Besides this many enzymes (acid phosphatase, nucleoside phosphorylase, RNA methylase etc. have been reported in the nucleolus.

Types of nucleoli : According to Wilson, there are two kinds of nucleous viz., Plasmosomes and Karyosomes (Ocata). The plasmosomes are positive to acidic stains and have a transparent outer region with a dense interior. The karyosomes stain positively with basic dyes. Three kinds of karyosomes have been identified-*Netknot* type (like nodules attached to threads), *Chromosome nucleoli* (occuring only in gametocytes), and *Karyospheres* (spheriodal bodies with basic chromatin).

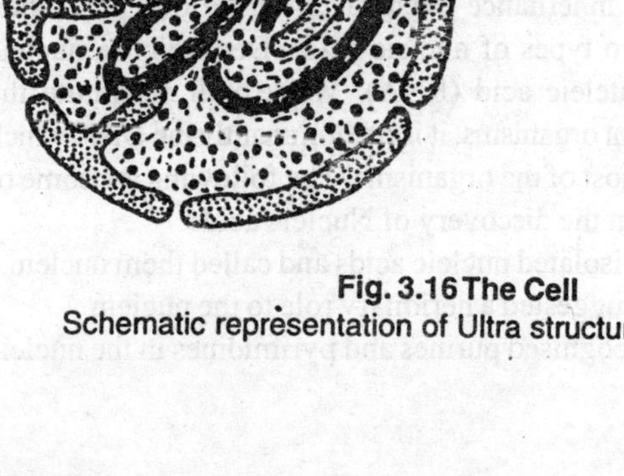

Fig. 3.16 The Cell
Schematic representation of Ultra structure of the nucleus

In some instances, plasmosomes and karyosomes combine to form amphinucleoli; these are common in molluscan eggs.

Fine structure of the nucleolus : Detailed analysis of the nucleolus has revealed the presence of three regions. These are granular region, fibrillar region and amorphous region. The granular, region occuring at the peripheral region of the nucleolus, is composed of granules of 150-250Å thick. This region has ribonucleic acid and proteins. The fibrillar region also known as nucleolonema is composed of fibrils 50-80Å in length. The amorphous region has low electron density and is the one to disappear first at the time of cell division.

There is no limiting membrane to the nucleolus and the organisation is is maintained by the calcium ions.

Functions of nucleolus

1. It is the active site of RNA synthesis
2. It is the source of ribosoinal RNA
3. It produces precursors of ribosomes
4. According to Maggio (1990), protein synthesis takes place in the nucleolus. But this may not be totally true.

Endosomes : These are granular structures present in the karyolymph and are smaller in size than the nucleolus.

Chemical composition of Nucleus : The main constituents of nucleus are nucleoprolcins, besides enzymes and inorganic salts. The following gives the total distribution of nucleoproteins in the mammalian liver cells–DNA 9%, RNA 1%, Histones 11%, Residual proteins 14%, and other acidic proteins 65%.

Nucleic Acids : These are the most important constituents of the nucleus playing a vital role in the inheritance of characters from one generation to-the other. There are two types of nucleic acids–Deoxyribose nucleic acid (DNA), and Ribose nucleic acid (RNA). While both constitute the genetic material in different organisms, it is predominantly the DNA which is the genetic material in most of the organisms. The following are some of the important milestones in the discovery of Nucleic acids.

1. Meischer (1869), isolated nucleic acids and called them nuclein.
2. Hertwig (1884), suggested a heriditary role to the nuclein.
3. Kossel (1884), recognised purines and pyrimidines in the nucleic acids.

4. Feulgen (1912), devised basic fuchsin dye positive only to DNA.
5. Astbury and Bell (1938), showed that the bases in DNA molecule are arranged one above the other.
6. Avery, Macleod and Macarty (1944), presented evidence for DNA to be the genetic material.
7. Chargaff (1950), showed that DNA contains equal amount of puriness and pyrimidines.
8. Wilkins et al (1950), identified the molecular distances between the bases.
9. Furberg (1952), opined that DNA is a coil of single chain.
10. Pauling and Corey (1952), suggested that DNA is a helix of three chains.
11. Waston and Crick (1953), demonstrated the double helix model for DNA molecule.
12. Meselson and Stahl (1958), demonstrated the semi-conservative replication of DNA molecule.
13. Jacob and Monad (1961), postulated the presence of RNA in protein

Deoxy Ribose nucleic Acid (DNA) DNA is the main genetic material in almost all organisms except for certain Viruses.

In prokaryotes, like *Escherichia* coil, the genetic material is a single giant molecule of DNA. While DNA is mainly concentrated in the nucleus, it has also been reported in cytoplasmic organelles like chloroplast, mitochondria etc.

Chemical composition of DNA : DNA extract from many living organisms have revealed it to be a complex macromolecule. It is a long polymer chain composed of monomers (units), called *nucleotides.* Each nucleotide in turn is composed of three sub-units. These are a) nitrogen base, b) sugar and c) phosphate.

Nitrogen bases : The nitrogen bases are of two kids viz., *pyrimidines* and *purines.* Both pyrimidines and purines are heterocyclic compounds, with a basic six membered ring in which one or the other carbon position is replaced by nitrogen.

The pyrimidines have a basic six membered ring of benzene in which, one or the other carbon position is replaced by nitrogen. There are two kinds of pyrimidines–*cytosine* and *thymine.* Structurally, cytosine has a single six membered ring with a hydroxyl group at the 2' position and an amino group at the 6' position (*2 hydroxy, 6 amino pyrimidine*). Thymine also has a single six membered ring with hydroxyl groups at the 2' and 6' position .and a methyl group attached to the 5' position (*2.6' hydroxy, 5 methyl pyrimidine*).

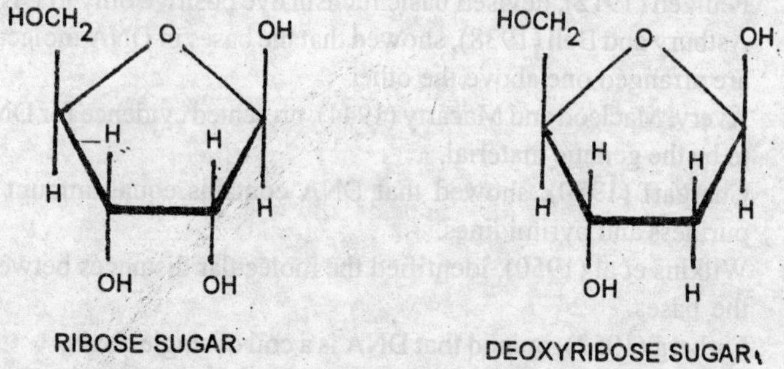

Fig.3.17 The Cell
Pentose sugars of nucleic acids (left, robose; right deoxyribose)

The purines have a double ringed structure. Attached to the six membered benzene ring is a 5 membered imidazole ring. The imidazole ring is joined to the hexagonal ring at 4' and 5' positions. The benzene ring has nitrogen in the 1' and 3' position (replacing C).

There are two types of purines viz., *adenine* and *Guanine.* Adenine has an amino group at the 6' position (*6 aminopurine*), while guanine has an aminogroup at the 2' position and a hydroxyl group at the 6' position (*2 amino, 6 hydroxy purine*).

Sugar : The sugar molecule in the DNA is basically a ribose sugar. The ribose sugar is a pentose sugar. It is a 5 membered ring. The ring consists of four carbon atoms while the 5th C atom is outside the ring. Position 1 '3' and 5', in the ring have OH groups while the 2nd carbon position has only H. There is no O; hence the sugar is said to be Deoxyribose (no oxygen in the second carbon position).

Phosphate : It has one central phosphorus with four oxygen atoms.
Nucleosides : A nitrogen base coupled with sugars forms a nucleoside. DNA has four nucleosides : a) Deoxycytidine, b) Deoxythymidine, c) Deoxyadenosine and d) deoxyguanosine. The sugar molecule is attached to the nitrogen base to the 3' position of the imidazole ring in the case of purines.

Nucleotides : A nucleotide is derived from the nucleoside by the addition of phosphoric acid molecule to the latter. The phosphate group is usually linked to the two sugar molecules above and below. The linkage is with the 3rd C atom of one deoxyribose sugar and with the 5th C atom of another.

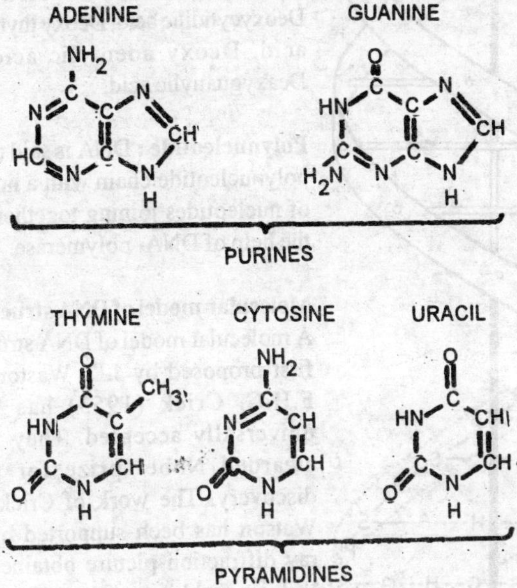

Fig. 3.18 The Cell
Nirtogen bases of nucleic acids

Fig. 3.19 The Cell
F.H.C. Crick (Left) and J.D. Watson (Right)

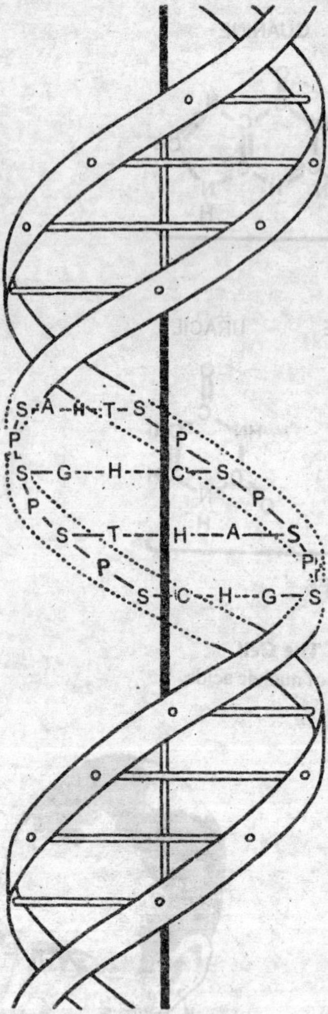

Fig. 3.20 The Cell
The double helix model of DNA

DNA has four nucleotides- Deoxycytidilic acid, Deoxy thymidilic acid, Deoxy adeniylic acid and Deaxyguanylic acid.

Polynucleotide : DNA is said to be a polynucleotide chain with a number of nucleotides joining together with the help of DNA–polymerase.

Molecular model of DNA structure : A molecular model of DNA structure first proposed by J.D. Waston and F.H.C. Crick (1953) has been universally accepted. They were awarded Nobel prize for their discovery. The work of Crick and Watson has been supported by X-ray diffraction picture obtained by Wilkins and his associates.

According to this model, a DNA molecule is composed of two chains that are spirally coiled and form a double helix, much like a spiral stairway. The distance between the two chains is uniform and it is maintained by base pairing. The back bone of the chain is made up of alternating sugar and phosphate molecules. At specific , intervals, the chains are joined by 'steps.' Each, step is made up of two nitrogen bases. Each nitrogen base is attached to the sugar molecule of the opposite chain. Between themselves, the nitrogen bases of a 'step' are joined by. of the nitrogen bases is always specific. It is made up of a purine and a pyrimidine i.e., a double ring and a single ring. Two purines would occupy too much space and can not be accommodated between two chains: similarly two pyrimidines would form a short step and leave a gap between two chains. Hence, the pairing is always between a purine and pyrimidine. Because of this pairing, in a DNA, molecule total purine content is always equivalent to total pyrimidine content

The purine pyrimidine base pairing is further specific. It is always Adenine with Thymine and Guanine with Cytosine. Adenine and Thymine are joined by two hydrogen bonds through atoms attached to positions 6' and 1'. Cytosine and Guanine are joined by three hydrogen bonds at position 6', I' and 2'. The bonding of hydrogen is weak; but this facilities easy separation of two strands during replication. According to Crick and Waston, the helix has a diameter of 20Å, and there is one complete turn of the helix at every 34Å, and has a stack of 10 nucleotides at every turn.

One unique character of DNA is its unity and yet infinitesimal variety. The DNA structure mentioned above is uniform in all organisms, whether it is man or virus. At the same time, it is also true that DNA is the genetic material that decides the characters of individual organisms. Then how does DNA decide the individuality. The answer lies in the base pair sequence. The combination of A/T, G/C is uniform, but their sequence is not; for instance in
one individual it may be AT, GC, AT, AT, AT, in another–GC, GC, AT, GC, AT etc. There is no limit to the variety in which base pair sequences are arranged. This is the secret of individual differences. In no two individuals, the base pair sequence is the same.

Unusual bases In DNA : The normal bases in DNA are A, T, G, and C. In some viruses like PBSIand PBS2 uracil replaces thymine. In some bacteriophages, C is replaced by 5 hydroxy methyl cytosine.

Single strand DNA : Although DNA in most organisms is double stranded in some phage viruses (fx 174), single stranded DNA is found. The following table summarises the differences between double stranded DNA and single stranded DNA

Table 3

	Character	Double Strand DNA	Single Strand DNA
1.	U.V. absorption	constant from 10-80°C and then rises rapidly	steady increase from 2° to 90°
2.	Action of formaldehyde	Resistant	not resistant
3.	Ratio of nucleotides	Equal, A = T and G = UC	A: T: G: C is 1.33:0:98:0.75

4. Shape	Linear	Circular
5. Replication	two strands separate and replication begins	becomes double stranded (replicate form) temporarily)

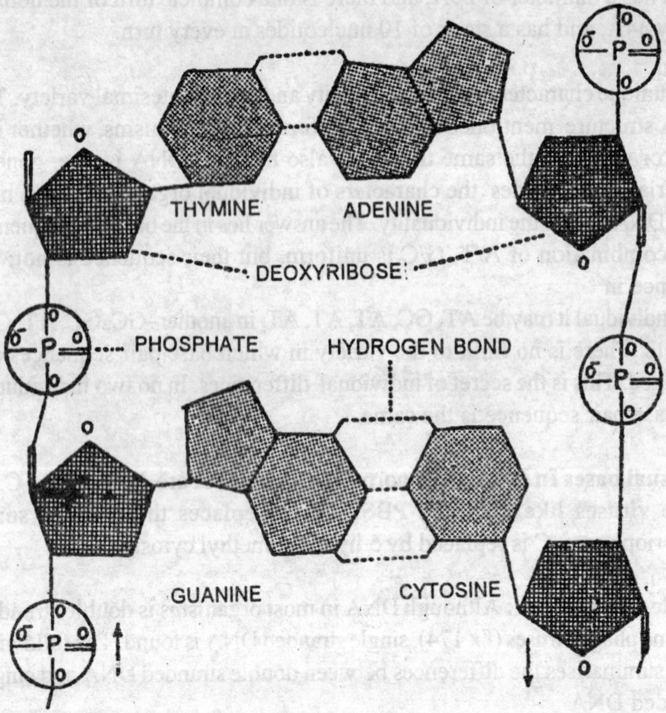

Fig. 3.21 The Cell

Arrangement of nucleotides in a segment of the DNA model

Ribonucleic acid (RNA) : The RNA is also a polymer of nucleotides. These are called ribonucleotides. Each ribonucleotide has three parts viz., a pentose sugar, a nitrogen base and a phosphate group.

The pentose sugar is a ribose sugar much like that of DNA, but with oxygen (present) attached to the 2' carbon position of the ring.
The nitrogen bases are– Adenine and guanine (purines) and cytosine and uracil, (pyrimidines).

Molecular structure of RNA : Here also, the back bone is made up of alternating sugar and phosphate. The single chain folds to enable the base pairs to meet here and there to form the steps. The pairing is between A-U and C-G. Occasionally the RNA occurs as the genetic material in *reoviruses.*

Types of RNA : The following are the types of RNA found – a) Genetic RNA b) Ribosomal RNA c) messenger RNA d) soluble or transfer RNA.

Genetic RNA : In certain viruses, RNA is the only genetic material, DNA being absent. This RNA is self-replicating; it is called RNA dependant RNA synthesis. The viral RNA directly functions as a messenger and produces RNA polymerase enzyme as well as proteins.

Ribosomal RNA (rRNA) : This is found in the ribosomes. It constitutes about 80% of total RNA of the cell. The base pairs of rRNA are complementary to that part of the nuclear DNA from where they are produced. rRNA consists of a single strand folding upon itself at some regions. The folded regions have paired bases, while the other regions have unpaired bases. Due to this in rRNA there is no purine pyrimidine equality.

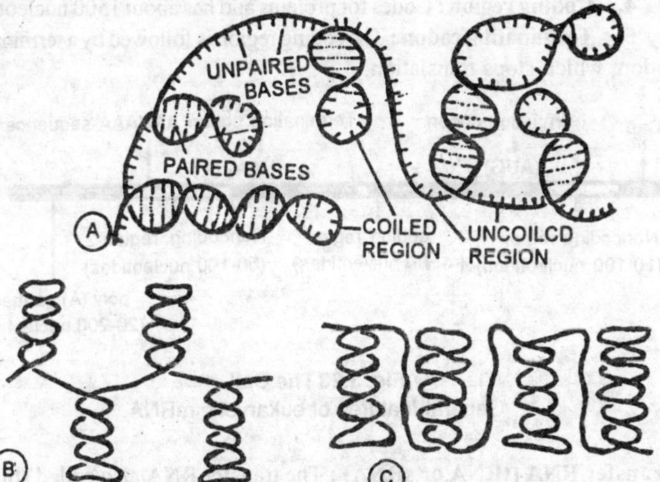

Fig.3.22 The Cell
Secondary structure of ribosomal RNA (Diagrammatic)
A. Extended strand, **B.** Negative interaction showing repulsion,
C. Positive interaction showing association

There are three types of rRNA based on sedimentation and molecular weight. These are a) rRNA with molecular weight over a million, eg. 21s – 29s rRNA, b) rRNA with molecular weight less than a million, eg. 12s–18s RNA, c) rRNA with very low molecular weight (40,000) eg. 5s rRNA.

Messenger RNA (mRNA) : mRNA carries the message from DNA to ribosomes during protein syntheses. It is about 3 – 5% of total RNA content of the cell.

mRNA has a molecular weight of about 5,00,000 with a sedimentation coefficient of 8s. This however, is highly variable, mRNA is always single stranded and there is no base pairing, even though random coiling may be observed. The mRNA molecule is produced from the coding strand and its base sequence is complementary to the coding strand. The following are some of the general features of mRNA.

1. **Cap :** This is a blocked methylated structure found at the 5' end of RNA.
2. **Non-coding region :** This follows the cap and consists of 10–100 nucleotides. There are no codes for proteins in this region.
3. **Initiating codon :** This region begins the coding.
4. **Coding region :** Codes for proteins and has about 1500 nucleotides.
5. **Terminating codon :** The coding region is followed by a terminating codon, which stops translation.

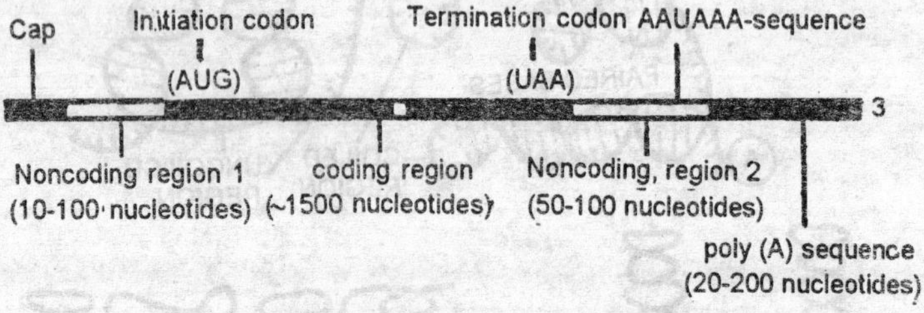

Fig. 3.23 The Cell
General features of eukaryotic mRNA

Transfer RNA (tRNA or sRNA) : The transfer RNA also called soluble RNA, is next in quantity to rRNA. It has a low molecular weight of about 25,000 – 30,000 and the sedimentation coefficient is 80s. It is made up of about 70 – 90 nucleotides.

tRNA is synthesised in the nucleus on a DNA template. Only a minute fraction of DNA (0.025%) codes form tRNA. The tRNA molecule consists of a single strand looped around itself with the 3' end always terminating in a C-C-A sequence. Some bases are paired.

The function of tRNA is to carry aminoacids to mRNA during protein synthesis. Each amino acid couples with a specific tRNA. Since the protein molecule consist of of about 20 types of amino acids, there must be at least 20 types or tRNA. But there could be many more, since there are two tRNAS for every amino acid in several, cases.

Mitochondria: (Gr. *mito* = thread, *chondrion* = granule)

Popularly called 'power house' of cells, these granular or filamentous organelles are present in all eukaryotes. They are characterised by their special structure, and unique physiological function concerning intracellular energy transformation.

Historically, mitochondria were first observed by Altmann (1894), who called them *bioblasts.* The present name was coined by Benda (1897), who developed the crystal violet staining. Bensely and Hoerr (1934), first isolated the liver cell mitochondria. The discovery that mitochondria are seats of cellular respiration was made by Hogeboom *et al,* in 1948.

It is difficult to observe mitochondria in living cells due to their low refractive index, but they can be studied under phase contrast microscope after staining with a dilute solution of janus green. Mitochondria may be isolated from cells by centrifugation between 20,000 g –40.000 g.

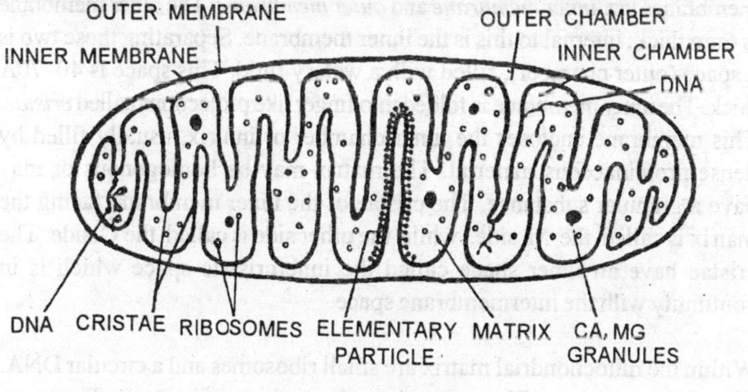

Fig. 3.24 The Cell

Diagrammatic representation of a section through a mitochondrion (ETPs are not shown on all christae)

Shape : Generally mitochondria are granular or filamentous, even through the shape is variable. During certain conditions, a long mitochondrion may swell at one end to become club shaped or look like a tennis racket .(De Robertis et al, 1970). Mitochondria may also become vesicular.

Size : Size also is variable. The width is about 0.5μm and the length may reach a maximum of 7μm. The size depends on the osmotic pressure.

Distribution : In general, mitochondria are distributed evenly in the cytoplasm, but in some instances they may accumulate around the nucleus or at the periphery of the cytoplasm. During cell division, mitochondria accumulate near the spindles. The distribution of mitochondria within the cell is in relation to the energy demand. For example in skeletal muscles they are grouped like rings or braces around the I-band of the myofibril.

Orientation : Mitochondria have more or less a specific orientation. In cylindrical cells, their long axis is parallel to the long axis of the cell. In leucocytes, mitochondria are arranged radially in respect to the centrioles.

Number : Mitochondria are absent in prokaryotic cells. The quantity of mitochondria per cell varies, but in liver cells, it is estimated that mitochondria contains 30–35% of the total protein content of the cell. A normal liver cell contains about 1000–1500 mitochondria. Some oocytes coniain as many as 300000 mitochondria.

Structure of mitochondria : The mitochondrion is enveloped by two membranes viz, *inner memhrane* and *outer membrame.* The outer membrane is 6nm thick; internal to this is the inner membrane. Separating these two is a space (outer chamber), filled wilh a watery fluid. This space is 40–70Å thick. The inner membrane is folded into finger like projections called *cristae.* This membrane encloses the inner chamber or innirix, usually filled by dense proteinaccous material. The matrix may be homogenous or may have a granular substance. The cristae of the inner membrane facing the matrix is called the M side, while the other side is called the C side. The cristae have an inner space called the innercristalc space which is in continuity wilh the intermembrane space.

Within the mitochondrial matrix are small ribosomes and a circular DNA. The ribosomes are of 70s type and similar to those of bacteria. These can synthesise proteins (Tyier 1973). The DNA called mDNA may be circular as in most of the animal cells or linear as in plant cells. Rabinowitch (1968), has demonstrated that the mitochondrial DNA has more G–C content than the nuclear DNA.

The mitochondrial DNA can replicate itself as it has DNA ploymerase. It is due to this, that mitochondria can replicate themselves.

The inner membrane : The inner membrane has numerous, nail like or tennis racquet like projections protruding into the matrix. Each particle is about 70–100A^0 in diameter. The particles are placed apart at a distance of 100A^0 and each is attached to the inner membrane by a short stalk,of 35–50A^0 long. The number of these particles per mitochondrion ranges from 10^4–10^5. These are called elementary particles, F_1 *particles* or *sub-units*. It was formerly believed that these particles contain all the enzymes necessary for electron transport and oxidative phosphorylation. Hence, they were named *electron transport particles* (ETP). Each particle was thought to have four complexes, with complex I and II in base, III in stalk and IV in the head piece. According to current thinking, the head piece has, ATPase, the stalk consists of a special protein (oligomycin sensitivity confering protein) and the piece contains the proton channel.

The inner membrane has the respiratory chain consisting of five complexes. Of these, four are involved in the electron transport, while the fifth one is involved in the synthesis of, ATP.

Mitochondria play a vital role in respiration. Glycolysis takes place in the *cytosol* (cytoplasm), while the enzymes of the Kerbs cycle are found in the matrix, those of oxidative phosphorylation are found in the inner membrane.

Biogensis of mitochondria : Each mitochondrion lasts for about a week or little more. There are three views regarding the origin of mitochondria–a) origin from cell membranes, b) Division of pre-existing mitochondria and c) *de novo* origin.

According to Robertson (1964), mitochondria in muscle cells arise by the infolding of the plasma membrane. This infolding joins another piece and forms a vesicular structure. In the nerve fibres of cray fish, endoplasmic reticulum may form mitochondria.

According io Luck and Rich (1964), mitochondria are capable of self-reproduction. DNA, protein synthesis, DNA polymerase and all the other essential, features for self-repliciation are present in mitochondria.

Harvey (1951) supports a *de novo origin* for mitochondria. Centrifugal separation of sea-urchin eggs into two-halves, a nucleated half (without mitochondria), and a non nucleated half (with mitochondira) showed that the nucleated half produced the mitochondria.

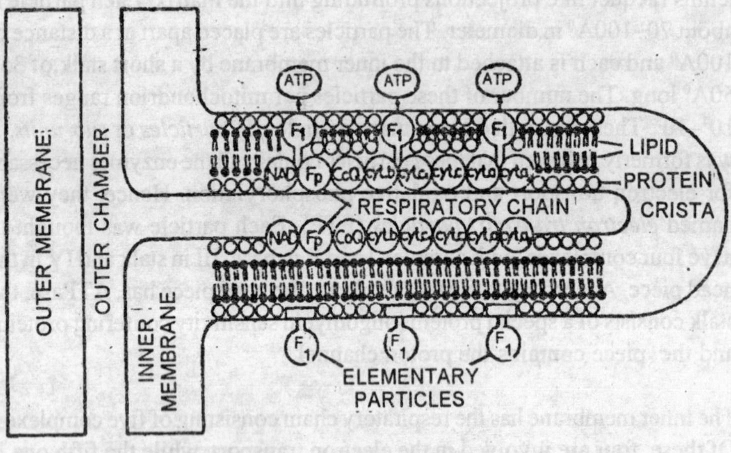

Fig. 3.25 The Cell
Arrangement of complexes (I-IV) in the ETP (old view)

Are mitochondria symbolic organisms? The many similarities between a bacterial cell and mitochondria made many people (Lloyd 1974; Borst 1977, O'Brien 1977 believe that the mitochondrion was once an autonomous entity which later entered into a symbiolic association with an anaerobically respiring host cell. The host cell derived its energy only from glycolysis. Subsequently, with the association, the host cell started respiring aerobically. De Robertis (1970) opines, that in evolutionary terms it is possible that a symbiotic relationship could have evolved into the present semiautonomus condition of the mitochondria.

The same theory has been proposed for chloroplasts also, as they also have all (the independent attributes of a mitochondria.

PLASTIDS : These are living cytoplasmic inclusions found in most of the plants except for some bacteria and fungi.

The plastids are of three categories viz., chromoplasts, leucoplasts and chloroplasts.

Chromoplasts : These are pigmented plastids. The pigments are non chlorophyllous like, carolenes, xanthophylls, fucoxathin, phycoerythrin etc. In photosynthetic bacteria, the chromoplasts have bacteropchlorophyll.

Leucoplasts : These are colourless plastids. They lack pigments and are usually present in cells which do not receive direct light. Leucoplasts may be seen in the storage leaves of onion. Leucoplasts that store starch are called *amyloplasts,* those that store oil are called elaioplasts and the ones storing proteins are called *aleuroneplasts.*

Chloroplasts : These are by far, the commonest and the most plastids. As the primary sites for trapping and converting solar energy they are very vital for the existence of not only the green plants, but for the whole living world.

Shape : Chloroplasts have varied shapes. The algal chloroplasts exhibit an array of shapes. They may be cup shaped *(Chlamydomonas),* girdle shaped *(Utothrix),* reticulate *(Hydrodictyon),* stellate *(Zygnema),* ribbon shaped *(Spirogyra)* or discoid. They are generally spherical or ovoid in higher plants.

Size : There is a great variety in size. Normally they are about lμm thick and 4–6μm in length. Chloroplasts of polyploid cells are generally larger than in the diploid cells. Cells in the shaded area have longer chloroplasts but with less intensity in colouration.

Distribition : They are uniformly distributed all over cytoplasm, but in some instances they cluster towards the nucleus. The concentration of chloroplasts will also depend on light intensity.

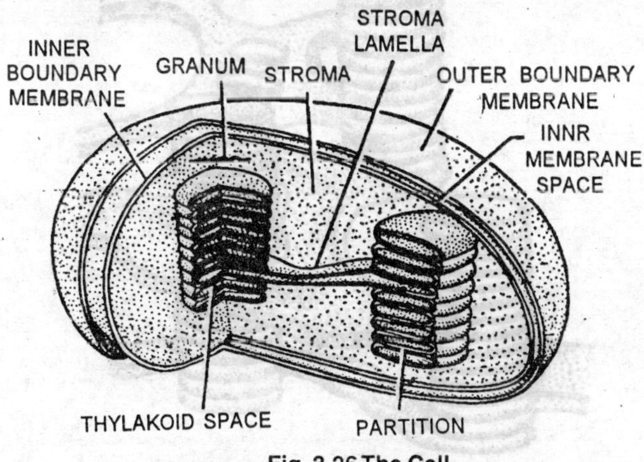

Fig. 3.26 The Cell
Model of Unltrastructure of chloroplasts

Number : The number of choloroplasts per cell varies. In some algae, like *Chlamydomonas,* there is only one chloroplast per cell or two as in *Zygnema.* In higher plants, cells have a number of chloroplasts. The leaf of *Ricinus communis* has nearly four lakh chloroplasts per square millimeter area.

Chemical composition : The chloroplasts contain carbohydrates, lipids, proteins chlorophyll, carotenoid and xanthophylls. The protein lipid ratio is 40:30. Besides the above, nucleic acids have also been demonstrated in the chloroplasts.

Ultra structure of the chloroplasts : The chloroplast has a covering of two membranes with an inner membrane space. These membranes are smooth and there are no perforations or particles. The membranes are differentially permeable (Mudrak)

A section of the chloroplast reveals an intricate system of membranes enclosed in a granular matrix. These membranes are called lamellae and the surrounding matrix--the *stroma.* In a sectional view, the lamellae can be seen packed and these stacks are called *thylakoids.* In the chloroplasts of higher plants, the thylakoids themselves form highly compact bundles called *grana.* Some thylakoids of granum extend into the stroma and maintain contact with other *grana.* These are *called-Stroma thylakoid* or *stroma lamellae* or *intergrana.*

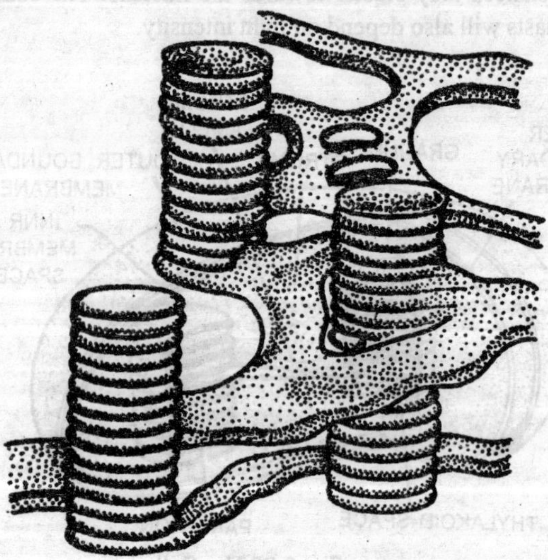

Fig. 3.27 The Cell
Three dimensional view of Grana (enlarged view)

The chloroplasts of algae lack the granum arrangement and in cyanobacteria (blue-green algae), the thylakoids lie naked in cytoplasm without any envelope. In such instances, pigments are uniformly distributed on or in the lamellae.

The *stroma* has in addition to the lamellae – granules (globuli), lipid droplets, starch grains and vesicles.

Structure of the lamellae : All the photosynthetic pigments are concentrated in the chloroplast lamellae. The lamellae as well as the outer membranes are composed of lipoprotein sub-units, each sub-unit according to Weir and Benson, has a protein core surrounded by a lipid sheath. The chloroplast membranes have a single layer of sub-units, whereas in thylakoids it is double due to the juxtaposition of the membranes. The appressed areas are called partitions. These are hydrophobic in nature The space between the two membranes of thylakoid (fret membrane), is called *fret channel.* The space between two thylakoids is called *loculus,* while the ends of disc shaped thylakoids are called *margins.* These areas (loculus, fret channel) are hydrophilic. The chlorophyll molecules are concentrated in the fret channel ie., in the space between the two membranes of thylakoids. The chlorophyll molecules may be Entirely in the space or the *heads* may be in the fret channel while the *tails* are buried in the sub-units, occasionally the entire chlorophyll molecule may be found in the fret membrane. Other pigments like cytochromes, carotenoids, etc., are also found in the fret membranes.

Four sub-units of the partition constitute a photosynthetic unit or a *quantosome.*

Ribosomes and RNA have also been isolated from the chloroplasts indicating a machinery for protein synthesis.

Since the chloroplasts, have their own RNA, DNA ribosomes and protein synthesis, according to Deviln (1975), "One might be inclined to view the chloroplast as a cell within a cell". De Robertis (1980) suggests, "It has bccn suggested that chloroplasts may have resulted from a symbiotic relationship between an autotrophic microorganism, able to transform energy from light, and a heterotrophic host cell" But Bogorad (1977) opines, that the chloroplast components depend to a great extent on the nuclear genes for their assembly into an integrated organelle.

Pigments in chloroplast : Some of the important pigments present in the chloroplasts are chlorophylls, carotenoids, cytochromes etc. In algal cells, the thylakoids also have phycobilins.

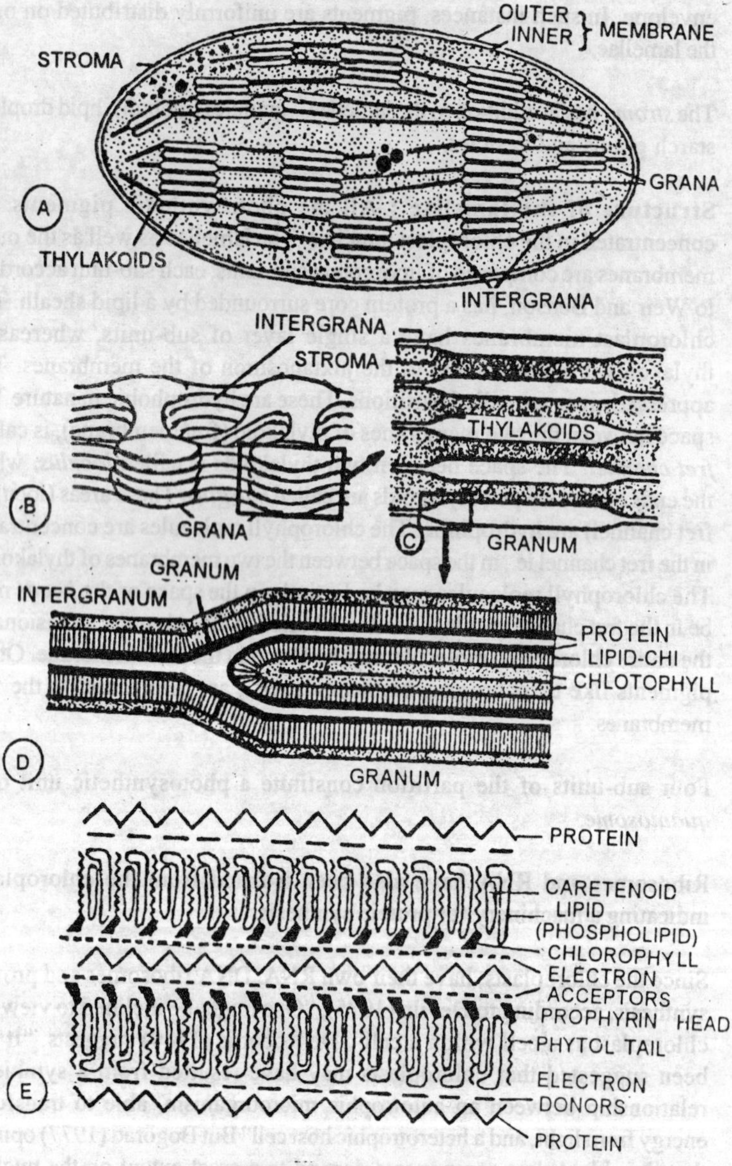

Fig. 3.28 The Cell
Cell structure of Chloroplast

Chlorophylls : These are the most important pigments involved in the basic photochemical reaction that sustains life. At least nine types of chlorophylls have been identified so far – Chlorophylls *a. b. c. d* and *e.* Bacterio Chlorophylls a and b and the chlorobium chlorophylls 650 and 660μm.

Among the chlorophylls, chlorophyll *a* and *b* are the best known and most widely distributed. These are absent in pigmented bacteria. Chlorophylls *c, d* and e are found only in algae in combination with chlorophyll a. Bacteriochlorophylls and chlorobium chlorophylls are found only in photosynthetic bacteria.

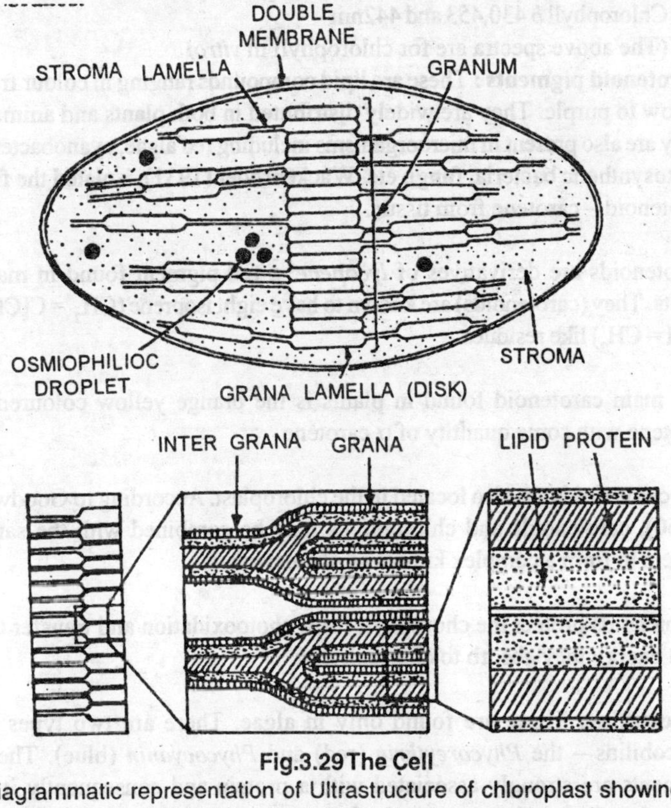

Fig.3.29 The Çell

Diagrammatic representation of Ultrastructure of chloroplast showing thelamellae. **A.** Entire chloroplast, **B-D.** Show successive enlargement of a part of the granum

The chlorophyll molecule is a complex structure basically made up of a 'head' and a 'tail' resembling a tennis racquet. The 'head' is a *porphyrin* structure made of four *Pyrrole rings* attached to each other at the centre by an isocyclic ring containing Mg atom at the centre. Extending from one of the pyrrole rings is the 'tail' – the alcoholic chain *(phytol)*. The emperical formula for the chlorophyll molecule is CH, O,N Mg. The phytol chain is esterified on the C atom of one of the pyrrole rings and has only one double bond.

Chlorophyll *a* and *b* differ structurally in having different atomic groupings at the C atom in the pyrrole ring. In chlorophyll a C atom has a methyl group, while chlorophyll *b* has an aldehyde group; besides the two pigments have different absorption spectra. The peaks are as follows –

Chlorophyll *a* 429, 410 and 660nm

Chlorophyll *b* 430,453 and 442nm

(The above spectra are for chlorophyll in *vitro)*

Carotenoid pigments : These are lipid compounds ranging in colour from yellow to purple. They are widely distributed in both plants and animals. They are also present in microorganisms including red algae, cyanobacteria, photosynthetic bacteria, fungi, etc. Wackenroder (1931), isolated the first carotenoid – carotene from tissue.

Carotenoids are derivatives of *lycopene,* a red pigment found in many plants. They (carotenoids) are known to have eight isoprene ($CH_2 = C[CH_3] - CH = CH_2$) like residues.

The main carotenoid found in plants is the orange yellow coloured β carotene with some quantity of α carotene.

The carotenoids are also located in the chloroplast. According to Goodwin (1960), carotenoids and chlorophylls may be combined with the same protein to form a complex known as *photosynthein.*

Carotenoids protect the chorophyll form photooxidation and transfer the light energy they absorb to chlorophyll *a.*

Phycobilins: These are found only in algae. There are two types of phycobilins – the *Phycoreythrin* (red) and *Phycocyanin* (blue). These pigments are strongly associated with a protein and consequently it is difficult to isolate the pigments in the pure state.

Like carotenoids, phycobilins are also involved in the transfer of light energy (they absorb) to the chlorophyll.

The absorption spectra of phycobilins presents anintersting study considering the fact that they act as accessory pigments in photosynthesis. R – phycoerythrin has peaks at 495, 540 and 545nm, while R – phycocyanin has peaks at 550 and 615nm.

ENDOPLASMIC RETICULUM : The cytoplasm in the cells of most of the plants and animals is traversed by a network of anastamosing fibrillar structures called endoplasmic reticulum (ER). Originally known as ergastoplasm, the current name ER was given by Porter and Kaallman (1952). The ER is connected to the outer nuclear membrane with the EF space opening into the perinuclear space between two membranes. The ER is known to be connected with the plasma membrane.

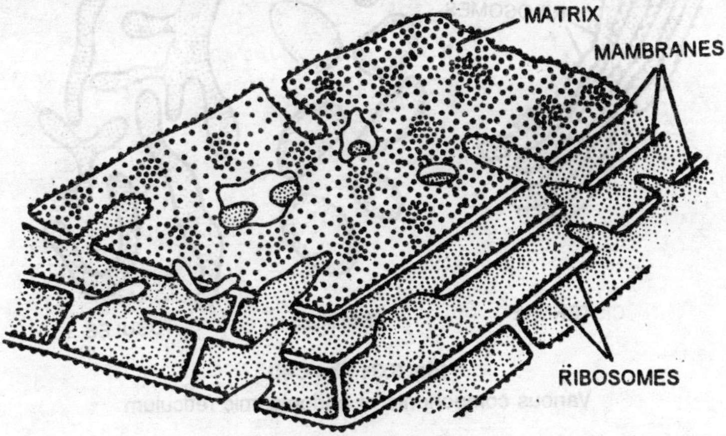

Fig. 3.30 The Cell
Three dimentional view of endoplasmic reticulum

Morphologically, the ER consists of three types of structures viz., *Cistemae, Vesicles* and *Tublllues*.

The **Cisternae** *are* elongated cylindrical unbranched structures, arranged parallel to one another respresenting a lamellated structure. They are 40 – 50µm in diameter. Cisternae are generally found in cells active in synthesis.

The **Vesicles** or sacs or oval bodies, attached to the membrane and have a diameter of 25-500µm. These are abundant in pancreatic cells.

The **Tubulues** are branched irregularly and are of diverse shapes having a size of 50 - 100µm. The tubulues are abundant in cells which are active in the synthesis of steroid compounds. They are also found in the epithelical cells of retina.

Ultra structure : All the three components of ER are bounded by a unit membrane 50 – 60Å thick. The membrane is trilamenar like the other unit membranes. Palade (1956) observed some secretory granules in the cavity of endoplasmic reticulum. The membrane has two outer layers of protein enclosing two layers of phospholipids. However, the ER membrane is symmetrical unlike in the other organelles (mitochondria).

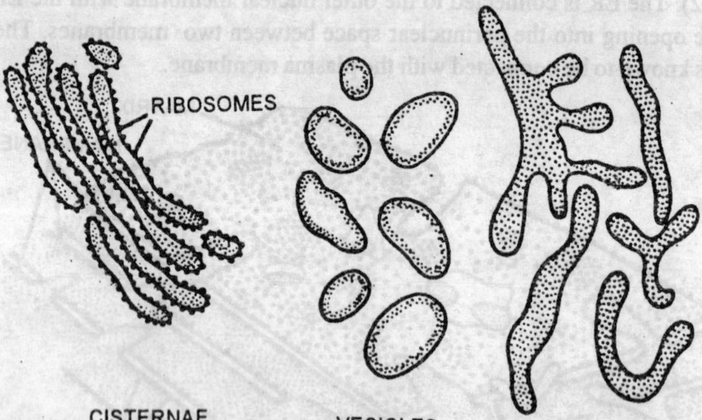

Fig. 3.31 The Cell
Various components of endoplasmic reticulum

Types of ER : Two types of ER have been observed in the cells (same or different). These are – smooth ER (agranular) and rough ER (granullar).

The membrane of smooth ER are not granular due to the absence of ribosomes. Smooth ER in mainly seen in tubules. The tubulues measure about 500 – 1000Å in length. Agranular ER is seen in cells involved in lipid or steroid synthesis. Cells active in protein synthesis do not have agranular ER. Agranular reticulum of muscle cells is called *sarcoplasmic reticulum* and it forms lace like sleeve around myofibrils. In the pigmented cells of retina, the agranular ER occurs in the form of tightly packed tubulues and vesicles called *myeloid bodies*.

The rough ER has a granular surface due to the presence of a large number of ribosomes attached to it. Since ribosomes are actively involved in protein synthesis, rough ER is conspicuous in cells that are actively involved in protein synthesis e.g. meristematic cells, plasma cells, goblet cells etc. The rough ER stains positively by basic dyes.

The cells (Spermatocytes) of certain vertebrates show an unnsual type of ER. Usually ER membrane does not have pores or openings. But in the instance mentioned above, pores or annuli have been reported resembling the ones found in the nuclear membrane. There is a diaphragm across the octogonal symmetry (Manul 1968).

The ER performs many vital cellular functions. These are –

(a) Enzymatic activities
(b) Transport of synthetic products
(c) Storage of metabolities
(d) Formation of nuclear envelope
(e) Intracellular impulse conduction etc.

Microsomes : These are structures that result due to the homogenization of the membranes. The cell membranes break up into fragments, round off and form microsomes. Microsomes are actually fragments of ER. They can be isolated by high speed centrifugation (100,000g) when they settle after nuclear and ribosomal fractions. Observed under electron microscope, they appear as membrane-bound vesicles of 500- 1500 Å in diameter. It should be understood however, that microsomes are not natural structures found in the intact cell. They are the result of homogenization of cell. The microsome fraction contains fragmented membranes.

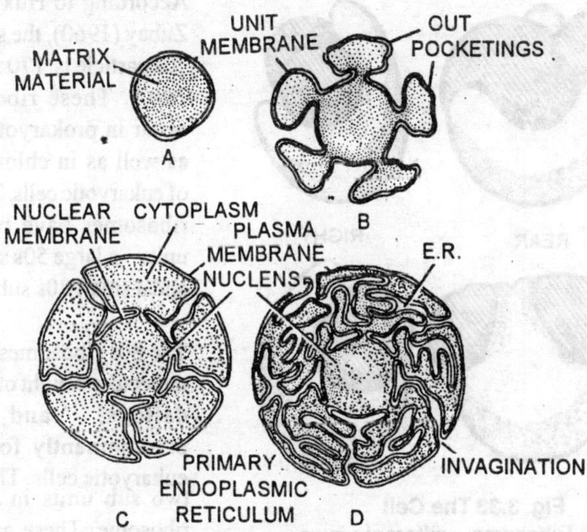

Fig. 3.32 The Cell

Stages in the origin of Endoplasmic reticulum

Origin of ER : The origin is uncertain, invagination of the plasma membrane might give rise to a canalicular system. But evidences for this hypothesis are yet to come (Dallmes et al,1966).

RIBOSOMES : These are dense granular nuclecoprotein structures occuring in cytoplasm, matrix of mitochondria and chloroplasts. In many instances while studying bean roots, ribosomes were later discovered in animal cells also. Ranging in diameter from 150 – 200Å, they have RNA and protein in equal quantities.

Ribosomes are isolated by differential centrifugation depending on sedimentation coefficient. The sedimentation coefficient is expressed in terms of *svedberg units.* The S units are related to the size and weight of the ribosome molecules.

Types of ribosomes : Two types of ribosomes have been identified based on the sedimentation coefficient which shows how soon a cell organelle can sediment under centrifugation. If the organelle is heavier, its sedimentation coefficient is more. The two types of ribosomes are – *70s ribosomes* and *80s ribosomes.*

The 70s ribosomes are comparatively smaller and have a molecular weight of 2.7×10^6 daltons (Dalton is a unit of molecular weight; one dalton = weight of one H atom = 1) According to Huxley and Zubay (1960), the size of a 70s particle is 170 x 170 x 200Å. These ribosomes occur in prokaryotic cells as well as in chloroplasts of eukaryotic cells. The 70s ribosomes have two sub units – a large 50s sub-unit and a small 30s sub-unit.

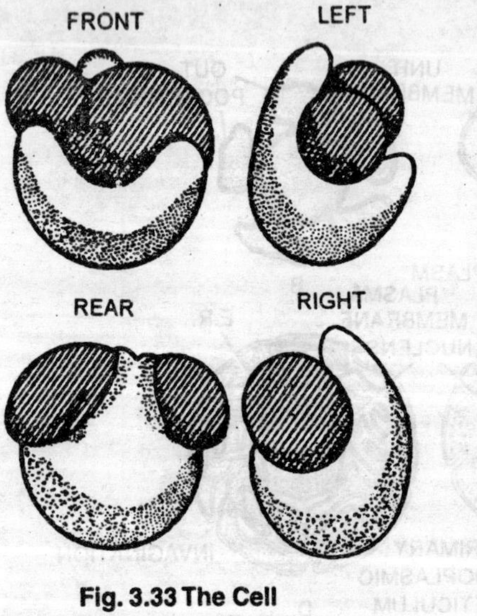

Fig. 3.33 The Cell
Ribosome - different views

The 80s ribosomes have a moleculer weight of 4×10^6 daltons, and are predominantly found in eukaryotic cells. There are two sub units in an 80s ribosome. These are – 60 and 40s.

Ribosomes may occur singly, as isolated units, and are called *monosomes*. When they occur in clusters or groups, they are called *polyribosomes*. *Tbe* polyribosomes may have a sedimentation coefficient of 100s – 600s.

The number of ribosomes per cell varies, it may be 10,000 (bacterial cell), or upto 10 million (eukaryotic cell).

Ribosomes of chloroplasts and mitochondria have their own protein synthesis. They have a sedimentation coefficient of 55s with two sub-units 40s and 30s.

Ultra structure of ribosomes : Ribosomes are oblate or spheroidal structures having two sub units (a large and a small). The large sub-unit is dome like and the smaller sub-unit is placed above like a cap. The 70s ribosomes have two units 50s and 30s as already mentioned. The 50s sub unit has a diameter of 140 - 160A^{0}and is pentagonal in shape. In the center of this sub unit occurs a round area of 40 - 60A^0. According to Fernando (1968), there is a pore like transparent area in the centre which prohibits the entry of proteolytic enzymes.

The pores have been found in the 50s sub-units of 80s ribosomes also. The smaller sub-units of both the types of ribosomes are often divided into two portions which remain connected to each other by a strand of 30 - 69A^0 thickness.

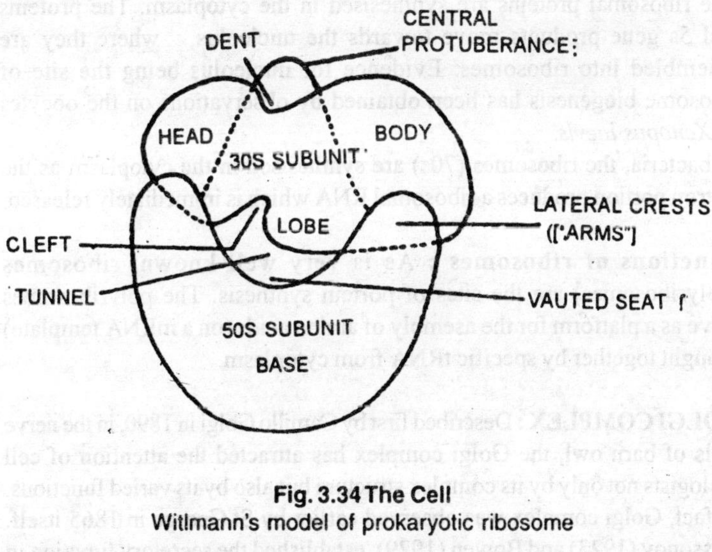

Fig. 3.34 The Cell
Witlmann's model of prokaryotic ribosome

Models for ribosomes structure : There are two models, viz., Wittmann's quasi symmetrical model and Lake's asymmetrical model.

According to Wittmann's model, the 30s sub unit is an elongate slightly bent stucture, having a head and a larger body with a cleft in between; with two parts protruding uniequally. It appears like a telephone receiver. The 50s sub unit is shaped variously – round shaped, maple leaf .shaped etc. The 50s sub unit has three protruberances.

The two sub units of 70s ribosomes have four areas of contact. The mRNA binding. region is located on the right side of the *30s* sub unit.
Lake's model regards the 30s sub unit to be totally asymmetrical. The sub unit is divided into two parts (1/3 and 2/3) unequally. Extending from the lower 2/3 region is a portion called the *platform*. This cleft is important in the sense of location of codon – anticodon interaction. The major difference between the Wittmanns model and Lake's model for the 30s sub unit is that the latter regards the 30s sub unit as totally asymmetrical. According to Lake, even the 50s sub-unit is asymmetrical.

Biogensis of ribosomes : The synthesis of eukaryotic ribosome is a complicated process involving several areas of the cell. The genes coding for 18s, 28s and 58s RNA are in the nucleolar organizing part of the chromosomes; the 5s genes are localized eleswhere in the chromosome. The ribosomal proteins are synthesised in the cytoplasm. The proteins and 5s gene products move towards the nucleolus, where they are assembled into ribosomes. Evidence for nucleolus being the site of ribosome biogenesis has been obtained by observations on the oocytes of *Xenopus laevis.*
In bacteria, the ribosomes (70s) are synthesised in the cytoplasm as the cistron portion produces a ribosomal RNA which is immediately released.

Functions of ribosomes : As is very well known, ribosomes (polyribosomes) are the sites of portein synthesis. The polyribosomes serve as a platform for the asembly of amino acids (on a mRNA template) brought together by specific tRNA from cytoplasm.

GOLGI COMPLEX : Described first by Camillo Golgi in 1890, in the nerve cells of barn owl, the Golgi complex has attracted the attention of cell biologists not only by its complex structure but also by its varied functions. In fact, Golgi complex was observed earlier by St George in 1865 itself. Nassonov (1923) and Bowen (1929), established the secretory function in the Golgi apparatus.

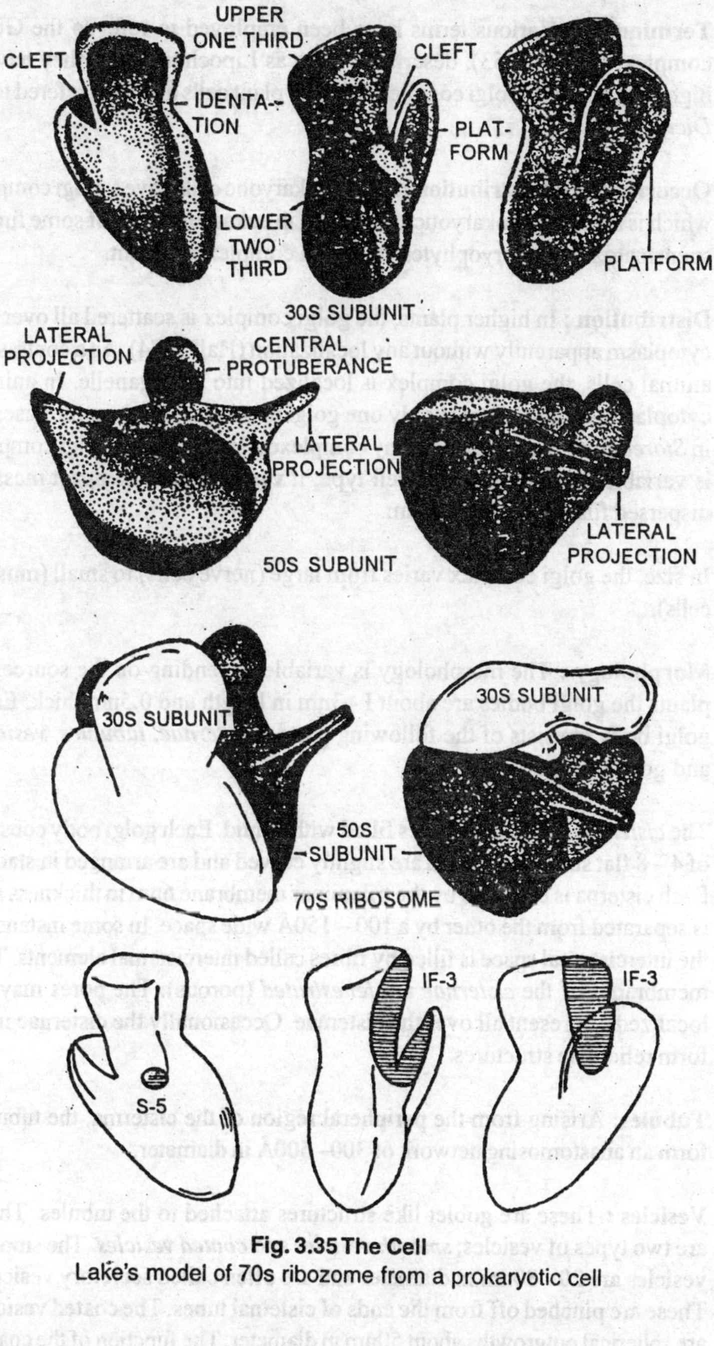

Fig. 3.35 The Cell
Lake's model of 70s ribozome from a prokaryotic cell

Terminoloy : Various terms have been employed to refer to the Golgi complex. Baker (1953), described, them as Lipochondria as they have a high lipid content. Golgi complex found in plant cells are often refered to as *Dictyosomes*.

Occurrence and distribution : All the eukaryotic cells have a golgi complex which is absent in prokaryotic cells. Even in eukaryotic cells of some fungi, spenrmatozoids of bryophytes etc. golgi complex is absent.

Distribution : In higher plants, the golgi complex is scattered all over the cytoplasm apparently without any localization (Hall 1974). As a contrast in animal cells, the golgi complex is localized into an organelle. In animal cytoplasm, usually there is only one golgi apparatus, but in some cases as in *Storeomyxa* there may be many complexes. In shape, the Golgi complex is variable depending on the cell type; it varies from a compact mass to dispersed filamentous reticulum.

In size, the golgi complex varies from large (nerve cells) to small (muscle cells).

Morphology : The morphology is variable depending on the source. In plants the golgi bodies are about I - 3nm in length and 0.5nm thick. Each golgi body consists of the following parts – *cisternae, tubulues, vesicles* and golgian vacuoles.

The *cisternae* are flattened sacs filled with a fluid. Each golgi body consists of 4 – 8 flat structures which are slightly curved and are arranged in stacks. Each cisterna is enclosed by the trilaminar membrane 6nm in thickness and is separated from the other by a 100 – 150Å wide space. In some instances, the intercisternal space is filled by fibres called intercistemal elements. The membranes of the *cisternae* are *fenestrated* (porous). The pores may be localized or present all over the cistemae. Occasionally the cisternae may form reticulate structures.

Tubules: Arising from the peripheral region of the cisterma, the tubules form an anastomosing network of 300 - 500Å in diameter.

Vesicles : These are goblet like structures attached to the tubules. There are two types of vesicles; *smooth vesicles* and *coated vesicles*. The smooth vesicles are 20 - 80nm in diameter and are often called secretory vesicles. These are pinched off from the ends of cistetnal tubes. The coated vesicles are spherical outgrowths about 50nm in diameter. The function of the coated vesicles is not known.

Vacuoles : These are large spaces occuring at the distal end of the cistermae. They represent the cisternae whose membranes are widely separated.

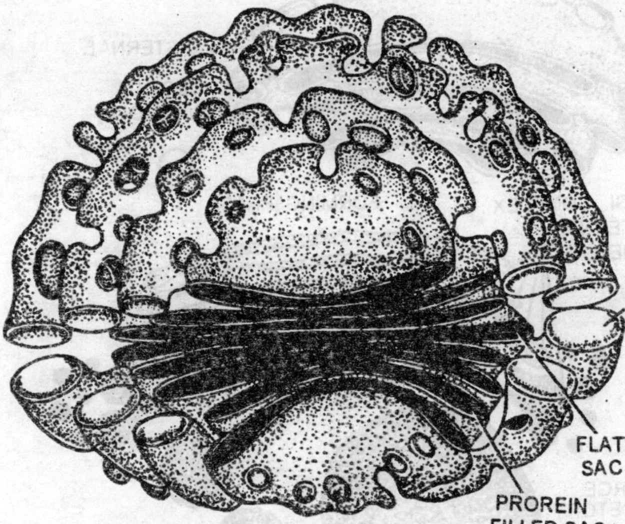

Fig. 3.36 The Cell
Golgi Complex

In the case of Dictyosomes (isolated golgi bodies) of plants, cisternae are extensively fesnestrated and have rough and smooth vesicles attached to them. Parallel fibres fill the intercistemal place.

Chemical composition of golgi complex : Proteins and phospholipids are in equal concentration. Golgi material in the nerve cells consists of cephalin and lecithin. Giroud has reported Vitamin C in the golgi complex.

Functions of the golgi complex : The following are some of the functions attributed to the golgi complex –

1. Absorption of compounds
2. Sites of enzyme production
3. Sites of hormonal production
4. Sites of protein storage
5. Forms acrosome during sperm maturation
6. Forms intracellular crystals

7. Formation of plant cell wall – the golgi bodies synthesise pectin, hemiceullulose and cellulose microfibrils. They also help in the formation of cell plate during mitosis.

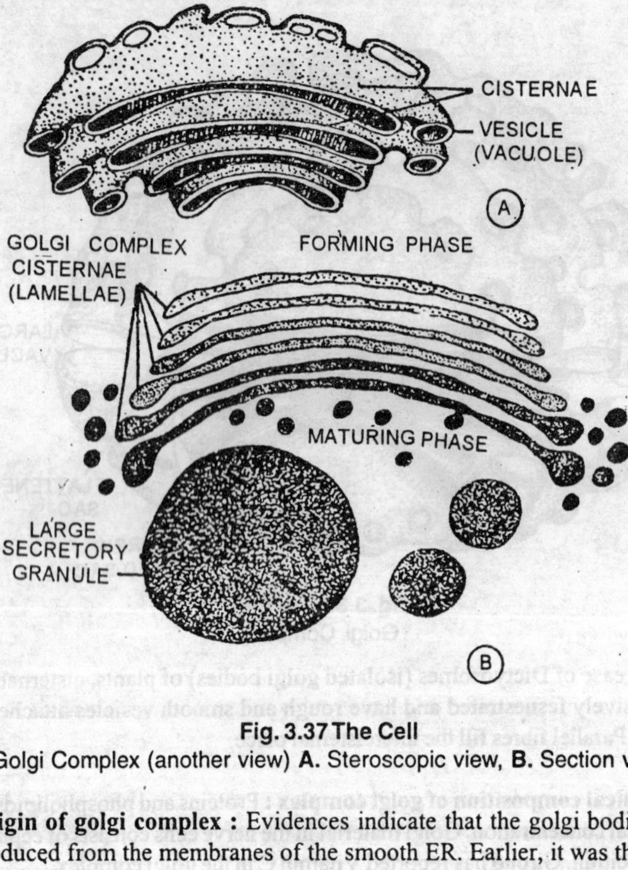

Fig. 3.37 The Cell

Golgi Complex (another view) **A.** Steroscopic view, **B.** Section view

Origin of golgi complex : Evidences indicate that the golgi bodies are produced from the membranes of the smooth ER. Earlier, it was thought that they arise *de novo,* or from the phragmoplasts of the cell wall. Both these theories are found to be incorrect.

Lysosomes : (*Gr.Lyso* = digestive, *soma* = body). First discovered in 1949 as dense round bodies, the lysosomes were called *pericanalicular bodies.* It was de Duve (1955), who gave the present name.

Occurence : Lysosomes are found in most of the animal cells and a few plant cells. The hepatic cells, pancreatic cells, kidney cells etc. possess large number of lysosomes. According to Matile (1969), the lysosomes in plant cell are membrane bound. Generally lysosomes are absent in the prokaryotic cells. Slime molds also have lysosomes.

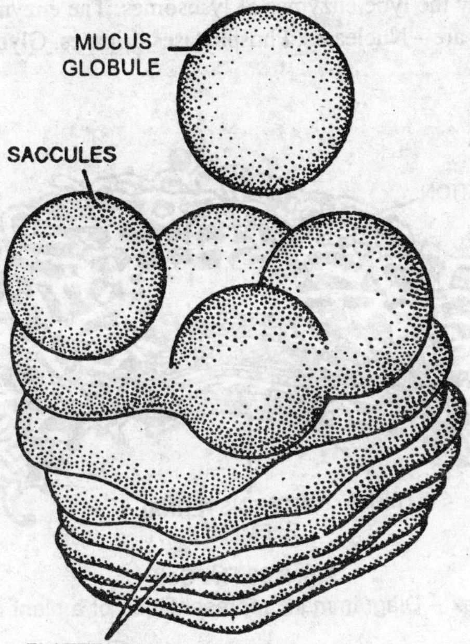

Fig. 3.38 The Cell
Golgi Complex with mucucus globules

Lysosomes are pleomorphic as seen in the meristematic cells of the root. The diameter may be 0.4 to 0.8μ; but in some mammalian kidney cells, lysosomes as long as 5μ have been found.

Ultra structure : Lysosomes are saccate structures filled with digestive enzymes. Each lysosome has two parts, viz., *limiting membrane* and *inner dense mass*.

The **limiting membrane** is not double membraned like in the other organelles. It is single and composed of lipoproteins. The membrane, however, is bilayered like any unit membrane.

The **inner dense mass** is the core of lysosomes; it is mostly enzymatic. The membrane is impermeable to substrates; the enzymes of which are present in the lysosomes. Occassionally certain substances called *labilizers* destabilize the membrane causing the enzymes to come out. Alternatively, there are substance called *stabilizers* which stabilize the membrane. The limited permeability of the membrane prevents excessive digestion of cellular components by the lytic enzymes of lysosomes. The enzymes present in the lysosomes are – Nucleases, Phosphatases, Lipases, Glycosidases, and Sulphatases.

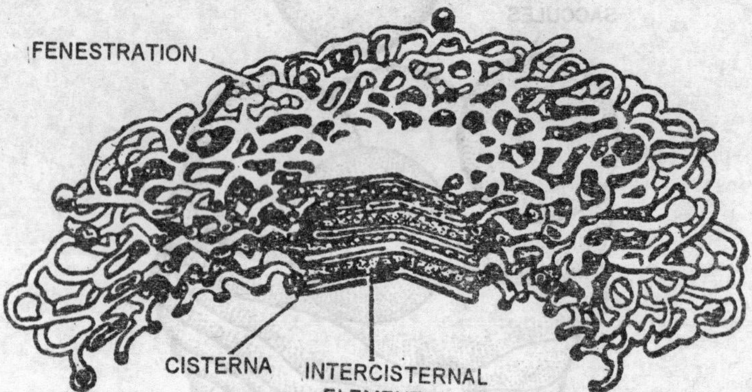

Fig. 3.39 The Cell
Golgi Complex – Diagrammatic representation of a plant dictyosome

Polymorphism : Lysosomes are ploymorphic, assuming various shapes depending on the cell type and situation. Basically there are two types of lysosomes viz., *Primary lysosomes and secondary lysosomes.* The secondary lysosomes are of three categories – *Phagosomes, residual bodies* and *autophagic vaculoes.*

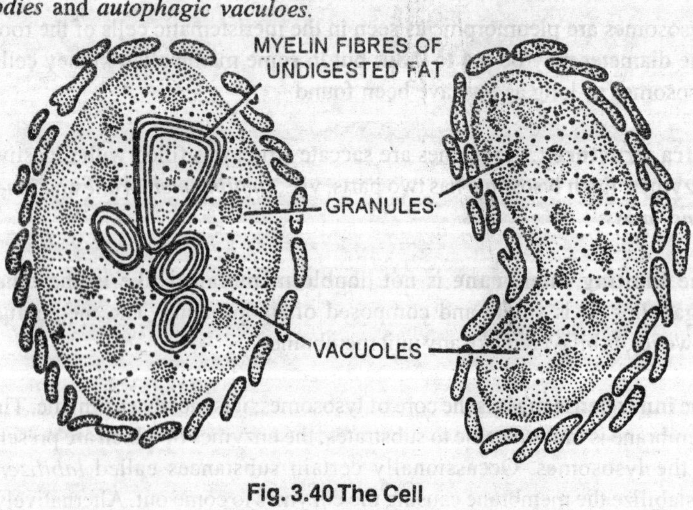

Fig. 3.40 The Cell
Lysosome – Structure

Primary lysosomes : These are sac like structures produced directly from ER from the Glogi bodies.

Phagosomes : These are also known as heterophagosomes or digestive vaculoes. When a foreign substance enters the cytoplasm, the plasma membrane forms a sac around these substances. These are the phagosomes or pinosomes. The lytic enzymes present in the phagosome digests the foreign substance. The phenomenon is called *Phagocytosis* or *pinocytosis*.

Residual bodies : After the digested material has passed out of the lysosomes, the undigested material remains within the lysososmes. Such lysosomes with residual substrate are called residual bodies. Residual bodies may remain within the cell or very often they are pushed out of the cell by exocytosis.

Autophagic vaculoe : Occasionally, a lysosome may digest a part of the cell (ER, mitochondria, etc.). This is a mechanism to digest a part of the cell without destroying the entire cell. Liver cells show autophagic vacuoles or autophagosomes under starvation.

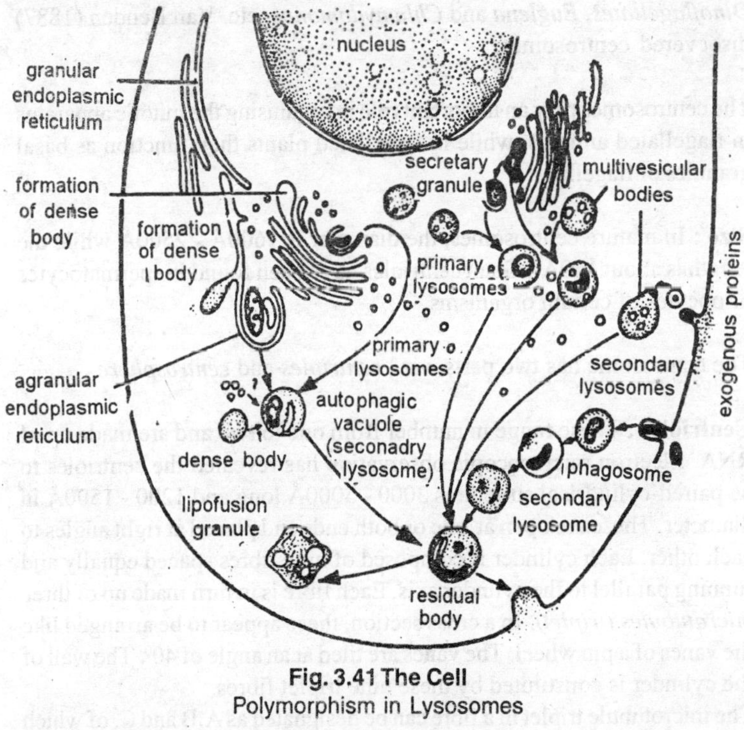

Fig. 3.41 The Cell
Polymorphism in Lysosomes

Origin of lysosomes

(a) *Extra cellular origin* : Vacuoles absorbed due to pinocytosis or phagocytosis may become secondary lysosomes after fusing with the primary lysosomes.

(b) *Origin from golgi bodies* : There are enough evidences to show that lysosomes are produced from golgi bodies at the height of their secretory activity.

(c) *Origin from ER* : According to Novikoff (1965), the rough ER gives rise to the lysosomes.

Major functions of lysosomes : The main function of the lysosomes is digestion of intracellular substances. According to Hirsch and Coh (1964), lysosomes help in destroying the dead cells from the tissue. Recently it has been found out that in some animals (frog), lysosomes play an important role in metamorphosis. Lysosomes, also perhaps help in the formation and destruction of bone cells.

CENTROSOMES : The centromes are found in the centre of the cell near the nucleus. These are present in all animal cells and lower plants like *Dinoflagellates*, *Euglena* and *Chlamydomonas,* etc. Van Benden (1887) disocvered centrosomes.

The centrosomes play an important role in organising the mitotic apparatus in flagellated animals, while in flagellated plants they function as basal granules of flagella.

Size : In mature centrosomes, the diameter is 1600Å - 2500Å while the length is about 160A^0. Giant centrioles have been found in spermatocytes or oocytes of certain organisms.

The centrosome has two parts, viz., *centrioles* and *centrosphere.*

Centrioles : These range in number from one to two and are made up of RNA. Electron microsocopic observation has revealed the centrioles to be paired-cylindrical structures 3000 - 5000Å long and 1200 - 1500Å in diameter. These are open at one or both ends and placed at right angles to each other. Each cylinder is composed of nine fibres spaced equally and running parallel to the cylinder axis. Each fibre is in turn made up of three *microtubules (triplet).* In a cross section, these appear to be arranged like the vanes of a pin wheel. The vanes are tiled at an angle of 40^0. The wall of the cylinder is constituted by these nine triplet fibres.
The microtubule triplet in a fibre can be designated as A,B and C, of which

A is innermost. Each microtubule wall is 50A^0 in diameter and the lumen in 120 - 130A^0 (the total diameter varies from 180 - 200A^0). The length of each microtubule is about 560A^0. At the innerside of each triplet fibre, a dense line runs and it is called *triplet base* connecting A and C tubules. In addition to this, a dense RNA region called *foot* is present.

Centrosphere : The dense granlar region of cytoplasm surrounding the centrioles is called the centrosphere.

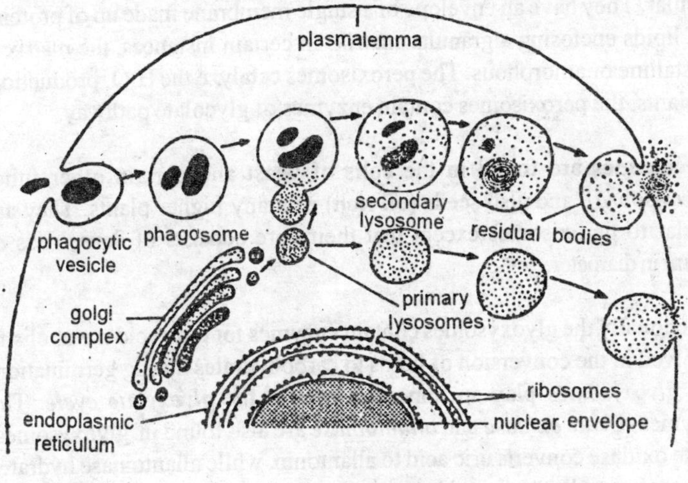

Fig. 3.42 The Cell
Origin and phases of Lysosomes

Structure : Structurally the microtubules are lipoproteinaceous (Fulton 1971).

Functions : Centrosomes appear to play an important part in cell division. people believe that centrioles play an important part in the formation of a spindle. They, however constitute the mitotic pole in higher animals. In primitive animals which are ciliated, centrioles seem to help in the generation of cilia.

Microbodies : The cells of plants, fungi, certain protozoa and liver cells of vertebrates contain certain small granular structures associated with ER, mitochondria and chloroplasts. These spherical bodies ranging in size from 0.2μ to 1.5μ, have a central crystalline core and are called *microbodies*.

Recent studies have revealed the presence of two types of microbodies viz., **Peroxisomes** and **Glyoxysomes.**

Peroxisomes are found in animal cells and also in the leaves of higher plants. These particles participate in the oxidation of substrates resulting in the formation of hydrogen peroxide. In plants, peroxisomes are involved in photorespiration.

Structurally peroxisomes are variable in shape and size but are generally circular. They have an envelope of a single membrane made up of proteins and lipids enclosing a granular matrix. In certain instances, the matrix is crystalline or amorphous. The peroxisomes catalyze the H_2O_2 production. In plants, the peroxisomes contain enzymes of glycolate pathway.

Glycosomes are found in the cells of yeast and certain other fungi *(Neurospora)*, and also seeds (oil rich) of many higher plants. They are similar to peroxisomes except that their core consists of dense rods of 6.0μm in diameter.

Functionally the glyoxysomes contain enzymes for fatty acid metabolism, involved in the conversion of lipids to carbohydrates during germination. The *glyoxysomes* play an important role in the *glyoxylate cycle.* The enzymes–*Urate oxidase* and *allantoinase* are also found in glyoxysomes. Urate oxidase converts uric acid to allantonin, while allantoinase hydrates allantonin to allantonic acid during the degrading of purines. (For non living inclusions of plant cell see College Botany Vol.II)

4

ENZYMES

A plant cell is virtually a microchemical factory in which hundreds of chemical reactions are taking place at amazingly high speeds and that at normal temperature ranges. These biochemical reactions provide sustenance for a plant to absorb, assimilate, synthesize, degrade, grow and reproduce which are essential for survival. Many of these reactions are simple, but most of them are complex, and involve several sequential steps in order to produce a particular product. Conversion of one compound to another has to be precise and exact, there being no excuse for mistakes. For a long time biologists and biochemists wondered, as to how biological reactions occur *in vivo* in such precise manner and at such rapid rates without involving undue increase of temperature. Very soon they got an answer, when Buchner (1987), accidentally discovered that a juice extracted from a yeast cell can perform fermentation as the living yeast cells can do. This means then that the juice contained some substance which can bring about fermentation. The chemical substance bringing about fermentation was soon isolated and given the name zymase. This started the glorious story of enzymes. Enzymes can be described as *Chemical middlemen* mediating and accelerating a variety of biological reactions. It may not be an exaggeration to state, virtually no metabolic activity in living organisms is participating without enzymes. Each reaction requires the participation one two or sometimes many specific enzymes. What are these enzymes? What is their role in metabolism? How do they accelerate chemical reactions? We shall try to find answers to these in this chapter.

The word enzyme is derived from Greek, meaning – *En* = in; zyme = leaven or living. The term was first coined by Kuhne (1978), while working on fermentation, even though it was Buchner who first made an extract of enzyme from yeast cells. Robert Sumer (1926), was the first scientist to purify and crystallize an enzyme revealing the proteinaceous nature. Even though the scientific discovery of enzymes may be comparatively recent, their use for practical purposes was known to early Greeks and Indians. The *Arthashastra* of Kautilya (4th century BC), mentions the use of extracts of barks in certain trees helpful in the process of fermentation.

Definition of enzymes

(a) Biocatalyst;

(b) Organic catalysts;

(c) Specific proteins, simple or compound in structure acting as specific catalyst;

(d) Biological middlemen.

Structure and chemical composition

All the known enzymes are proteinaceous. These proteins called functional proteins, have a very complex structure and have molecular weights sometimes ranging up to several millions. Even some of the simple enzymes like *pipsin* (35,000), *urease* (4,83,000) etc., also have high molecular weights. While all enzymes are made up of proteins, they also have a non-protein part. The bulk of the enzyme, however, is proteinaceous.

The total eyzyme is given the name *holoenzyme* (Protein + non protein), while the protein part is called the *apoenzyme* and the non-protein part is given the name *prosthetic* group. When the prosthetic group is not tightly bound to the apoenzyme, it is called a *coenzyme* or *cofactor*. The presence of a non-protein part in the enzyme is a must as otherwise the enzyme will be inactive. The following are some of the structural attributes of the enzymes.

(i) Apoenzyme

The protein part constitute the major part of the enzyme molecule. The complex proteins that make up the apoenzyme have specific sequences of aminoacids contributing to the specific reactivity of the enzymes. As the proteins are colloids, they offer a large area in relation to their volume. Very few enzymes are wholly made up of proteins. E.g. *proteolytic enzymes*.

Active site

All enzymes possess an area in their molecular organization where substrate materials bind themselves in order to undergo chemical change. This binding site is called the active site of an enzyme. An enzyme may have one or more active sites. In addition to the active sites an enzyme may also have *regulatory sites*, to which regulatory substances bind and regulate the activity of the enzyme.

Prosthetic group

Most of the enzymes possess a non-protein part tightly bound to the protein part known as the prosthetic group. Many metallic ions like Cu, Zn, Mo etc., constitute the prosthetic groups. Organic compounds such as cytochromes, flavoproteins, pyridoxal phosphate, biotin etc., are also known to be prosthetic groups of certain enzymes.

Coenzymes and Cofactors

Strictly speaking, there is no distinction between a prosthetic group and a coenzyme except that the latter is loosely bound to the enzyme molecule. The term cofactor is employed for inorganic ions such as Cl^- Mg^{++} etc., while organic molecules are called conezymes.

Distribution of enzymes in plant cells

Enzymes are found distributed all over the plant cell. But there is a qualitative distribution i.e., not all enzymes are found in all the areas. Enzymes have a localized distribution with reference to the functions they have to perform. The distribution of some important enzymes are listed below.

1. **Nucleus** : *Polymerases, Transferases* etc.
2. **Chloroplast** : *Carboxydismutases, Phosphatases, Kinases, Dehydrogenases* etc.
3. **Mitochondris** : Enzymes of respiratory pathway - *Dehydrogenases, Cytochrome oxidases etc.*
4. **Cytoplasm** : *Aldolsases, Isomerases, ATP ases, Phosphorylases, Transphosphorylases* etc.

(The list is not comprehensive)

Nomenclature and Classification

The name of the enzyme is usually descriptive and consists of two parts – the first part denotes the substrate on which the enzyme acts, while the second part indicates the type of action of enzyme. For example, *Isocitric dehydrogenase* indicates the enzyme acts on Isocitric acid and removes hydrogen from it. In some of the older nomenclature, this is not followed and the enzymes are named arbitarily. For Eg., *Pepsin, Trypsin, Chymotrypsin* etc.

According to the commission on Enzymology of the International Union of Biochemistry, all enzyme names should end with ase.

Enzymes are classified into six major divisions. These are –

1. Oxidoreductases

2. Transferases
3. Lyases
4. Isomerases
5. Ligases (= synthetases) and
6. Hydrolases.

1. Oxidoreductases

Enzymes which bring about oxidation reduction reactions are included under this group. This group is further subdivided into the following.

(i) *Oxidases* : Reactions involving oxidation of the substrate with molecular oxygen serving as the electron acceptor belng to this category.

(ii) *Peroxidases* : These catalyze the removal of hydrogen from the substrate which combines with hydrogen peroxide (H_2O_2)

$$SH_2 + H_2O \rightarrow \quad 2H_2O + S$$

(iii) *Dehydrogenases*: Enzymes which bring about oxidation of the substrate by removing hydrogen from it belong to this category. The hydrogen acceptor, however, is not molecular oxygen.

$$SH_2R \rightarrow \quad S + RH_2$$

(iv) *Reductases* : These catalyze the incorporation of both the atoms of oxygen into the substrate.

(v) *Oxygenases* : These catalyze the incorporation of both the atoms of oxygen into the substrate.

(vi) *Hydrolyases* : These enzymes catalyze the reaction in which only one atom of oxygen is added to the substrate.

(vii) *Catalases* : These liberate molecular oxygen from hydrogen peroxide.

2. Transferases

These enzymes transfer a particular group from one substrate to another. These are further divided as follows : based on the type of group transfer which they catalyze.

(i) *Transketolases* : Transfer a ketone group
(ii) *Transaminase* : Transfer an amino group
(iii) *Transaldolases* : Transfer an aldose group
(iv) *Transphosphorylases* : Transfer an every rich phosphate group
(v) *Transcarboxylases* : Transfer a carboxyl group
(vi) *Hexokinases* : Transfer an ordinary phosphate group
(vii) *Gransglycorsylases* : Transfer of a glycosyl group

3. Hydrolases

These enzymes catalyze the hydrolysis of complex substrates into simpler ones (starch into glucose). These include *Carbohydrases* (hydrolysis of carbohydrates), *Lipases* (hydrolysis of fats), *Phosphatases* (removal of phosphate groups) etc.

4. Lyases

These catalyze the removal of groups from the substrate without the addition of water (non hydrolytic).

Fructose, 1,6 diphosphate → Dihydroxy acetone phosphate +phosophoglyceraldehyde

Other examples of lyases are - *Aconitase, Isocitrase, Fumarase* etc.

5. Isomerases

They bring about isomerization of one substrate to another (intramolecular shifting of atoms)

$$\text{Glucose 6 phosphate} \xrightarrow{\text{Isomerase}} \text{Fructose 6 phosphate}$$

6. Ligases

Also called synthetses, they help in the synthesis of a new compound generally with the cleavage of ATP or other nucleoside phosphates.

Isoenzymes

Enzymes which perform similar functions but having different molecular structures are called *Isoenzymes* or *Isozymes*. In other words, *different molecular species of an enzyme can be called Isozymes.* Isozymes can be identified based on the appearance of different bands on a polyacrylamid 3 gel. For example, the enzyme *luctic dehydrogenase* has five isozymase.

Zymogens

These are enzyme precursors or inactivated forms of enzymes secreted by cells. These can be activated to the normal enzyme state by chemical modification.

Constitutive and adaptive enzymes

Not all enzymes are present in the site of action always. Some are synthesized at the time of reaction and soon get degraded after the reaction. Such enzymes are known as *adaptive enzymes*. The other category – *constitutive enzymes* are present always and participate in reaction whenever required.

Properties of enzymes (Physical and Chemical)

Enzymes are characterized by the following properties.

1. Catalytic properties

All enzymes are catalysts, and as such they accelerate the pace of reaction. Being true catalysts, they do not undergo any chemical change and do not figure in the end products of a reaction. They also do not disturb the equilibrium of a reaction. Being catalysts, they are required in very small quantities. The number of molecules of substrate converted into product per minute is called, the *turnover number* of the enzyme indicating its efficiency. Turnover number of different enzymes varies from 100-3,00,000.

The velocity of enzyme action like that of a true catalyst is proportional to the concentration of the substrate/concentration of the enzyme.

2. Reversibility of action

Like all catalysts, enzymes can accelerate the pace of a reaction in both directions. The direction depends on the factors present at that particular time. It should be specified here that an enzyme by itself does not initiate a reaction but merely speeds up the rate. Thus, an enzyme *starch phosphorylase* catalyzes the hydrolysis of starch into glucose during day, but brings about the synthesis of starch during darkness. So here, it is not the enzyme that decides the direction of a reaction, but the presence or absence of light, pH etc., which do so. The enzyme merely speeds up the rate of reaction in whichever direction it is proceeding.

3. Specificity

Enzymes are highly specific in their action. A particular enzyme catalyzes only a particular kind of a reaction. Sometimes, if a reaction involves several steps, an enzyme will be specific for each step. Enzyme specificity is of the following types.

(i) *Absolute specific* : Acts on only one kind of a substrate. For example, *malic dehydrogenase* acts only on malic acid.

(ii) *Absolute group specific* : Acts on a given organic group. For e.g., *alcoholic dehydrogenase* acts only an alcoholic group.

(iii) *Relative group specific* : Enzyme catalyses more than one organic group. E.g., *Trypsin* acting on ester bonds involving Arginine or lysine residues.

(iv) *Optical specificity* : Certain enzymes can distinguish between optical isomers in the substrate and catalyze any one only. For example -*amino acid oxidase* will not act on D - amino acids and *vice versa*.

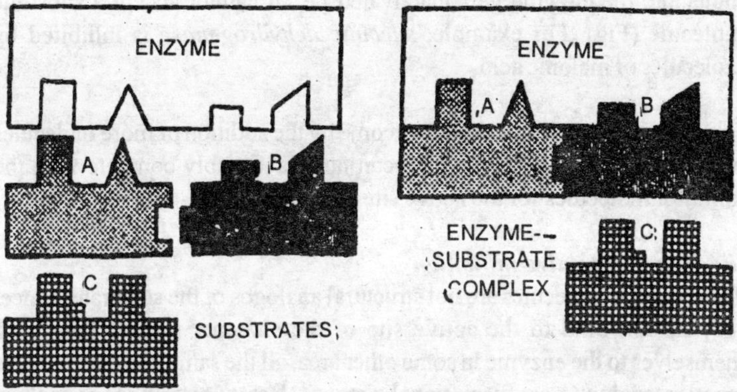

Fig. 4.1 Enzymes
Diagram to illustrate enzyme substrate specificity

4. Heat sensitivity

Enzymes remain active under normal temperature and are denatured under high temperatures (60 - 70^{0}C). Under dehydrated conditions however, as in seeds, they can withstand temperature up to 120^{0}C. Hence enzymes may be termed *thermolabile.*

5. pH sensitivity

Most of the enzymes are active only under a particular range of pH. Any alteration in pH would render the enzymes inactive. Enzymes are denatured by strong acids or alkali. Optimum pH for each enzyme varies depending on the nature of the buffer system (if present), absorbed ions, presence of activators/inhibitors etc.

6. Inhibition of enzyme action

Most of the enzymes have active sites to which the substrate binds to undergo conformational changes. In some instances, it has been noticed that certain other substances which are partially indentical to the substrates bind themselves to the enzyme and inactivate it. These substances are called inhibitors and their action is called inhibition. There are three types of inhibition *viz.* **competetive inhibition, non competetive inhibition** and **uncompetetive inhibition.**

(i) Competitive inhibition

In this, the inhibitor molecule is structurally analogous with the substrate molecule. It competes with the substrate molecule to occupy the active site of the enzyme. Due to the occupation of the active site by the inhibitor molecule, the enzyme gets inactivated i.e., it cannot accept a substrate molecule (Fig). For example, *succinic dehydrogenase* is inhibited by molecules of malonic acid.

Competitive inhibition can be overcome by the addition of more molecules of substrate to the system which can now favourably compete with the inhibitor molecules for the active site of the enzyme.

(ii) Non competitive inhibition

The inhibitor molecules are not structural analogos of the substrate, hence, they do not bind to the active site of the enzyme, instead they bind themselves to the enzyme in some other area, all the same, the enzyme gets inactivated due to conformational changes. Heavy metallic ions such as Mg ++, Ag++, Pb++ etc. combine with the sulphydryl groups of the enzyme rendering it inactive.

Non competitive inhibition is irreversible. Addition of extra molecules of substrate is of no consequence since the active sites are anyway available to the substrate.

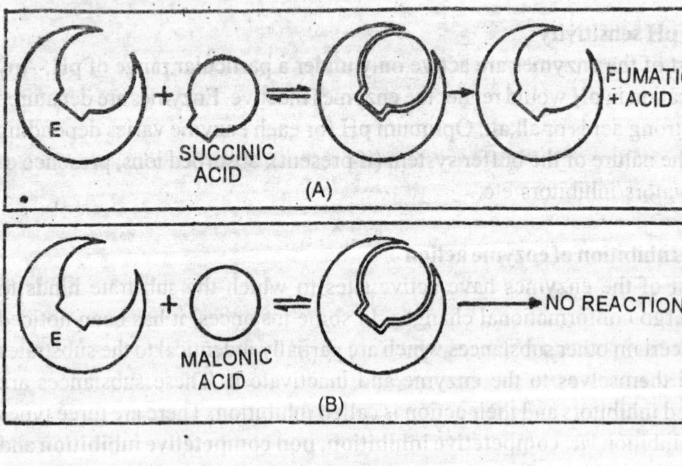

Fig. 4.2 Enzymes
Competetive inhibition of enzyme action

(iii) Uncompetetive inhibition

In competetive inhibition, the inhibitor is supposed to combine with forms of the enzyme but does not combine with the substrate enzyme complex. The form of the enzyme to which uncompetetive inhibitor attaches will always remain substrate non combining, in the sense it can never get back to the substrate combining form of the enzyme. As a result, enzymatic catalyses will stop. This inhibition also cannot be remedied by adding more amount of substrate. Uncompetetive inhibition is very common in multisubstrate reactions, but very rarely in single substrate reactions.

6. Colloidal condition

The enzymes are large molecules of proteins and exhibit all the colloidal properties. They cannot pass through a membrane of *collodion.* The enzymes have the capacity of absorption.

7. Redox potential value

Certain enzymatic reaction which are reversible, depend on the redox potential value for the direction of the reaction

MECHANISM OF ENZYME ACTION

How does an enzyme catalyze, or speed up a reaction? The following explanation may be offered by considering the conditions of a biochemical reaction.

Normally, chemical reaction take place when molecules of any substance possessing on an average a higher kinetic energy change into a substance whose molecules have a lower average of kinetic energy. Although on an average all molecules of reactions (substrate) possess equal amount of kinetic energy, a few of them possess higher kinetic energy, and a few of them possess lower than the average due to collisions among molecules. Thus among the molecules of a substrate, there are *energy rich* and *energy poor* molecules. If we, for instance take a reaction where molecules of A get converted into B, at any given time only few molecules of A can get converted into B because of their higher kinetic energy. This extra energy required for the participation in the reaction is called **activation energy.** The activation energy is always above the average kinetic energy of the molecules. In a non enzymatic reaction, the average kinetic energy of molecules may be raised to activation energy by increasing the temperature of the medium. This speeds up the rate of reaction. Theoretically, continuous increase of temperature should indefinitely increase the rate of reaction. But this is not possible actually because, a higher temperature beyond a certain point would kill the whole system. The enzymes in living cells do the same things in *vivo,* what the temperature does in *vitro* (they speed up the rate). But the enzymes do not increase the kinetic energy of the molecules. Instead, they combine with the molecules and lower the activation energy necessary for reaction. The combined product of enzyme and substrate known as *enzyme substrate complex,* crosses the activation energy barrier much easily than what the plain substrate molecules can do. All enzymatic reactions are believed to take place as follows.

(i) E + S → ES

(ii) ES → A + B + E

(E = enzyme; S = Substrate; ES = enzyme substrate complex; A and B = products)

Measurement of activation energy has been done for several reactions with or without enzymes. With the participation of enzymes, it has been noticed that more number of substrate particles can participate in a reaction than what is normally possible. For example, hydrolysis of sucrose with only hydrogen ions as catalyst has an activation energy of 25,000 cal/mol., but with the addition of yeast *invertase,* it comes down to 10,000 cal/mol. Similarly, in the decomposition of H_2O_2 (enzyme *catalase*), activation energy comes down from 18,000 cal/mol to 2000 cal/mol; for caesin the figures are 20,600 cal/mol and 12,000 cal/mol without, and with enzyme (*trypsin*).

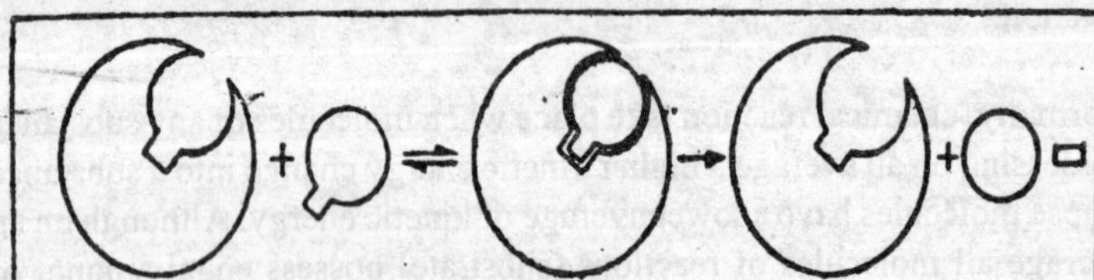

Fig. 4.3 Enzymes

Graph to demonstrate lowering of activation of energy due to enzyme catalysis

Enzyme kinetics (Michaelis - Menton equation) : Michaelis and Menton (1913), proposed a theory for enzyme action, in which they believed that the substrate and enzyme form a loose complex, and at the end of reaction the complex breaks releasing the enzyme and products as explained above. In order to explain the rate of substrate conversion, they applied the laws of kinetics which may be represented as follows.

K_1 (E = enzyme; S = substrate; ES = enzyme

E+S → ES substrate complex; K_1 velocity constant of the formation of ES and K_2 velocity constant of dissociation of ES)

OR

$$E + S \xrightarrow{K_1} ES \xrightarrow{K_2} P + E$$

(K_2 = velocity constant for decomposition of ES.P = end product)

The equation can be further simplified as $\frac{K_1 + K_2}{K_1} = K_M$

K_M is known as the velocity constant, or Michaelis and Menton's constant. Its value is characteristic for each enzyme and substrate and by comparing KM values of an enzyme with different substrates, the affinity of different substrates to an enzyme can be calculated. When the velocity of the reaction is half of the maximum velocity, KM is equal to the concentration of the substrate. However, Michael is Menton equation is applicable only to single substrate reactions.

MODELS FOR ENZYME ACTION

Two models have been proposed to account for the specificity and mode of enzyme action. These are (i) the lock and key theory, and (ii) the induced fit theory.

Lock and key model

This theory originally proposed by Fischer (1898), believes that the enzyme specificity can be accounted for by assuming that the enzyme and the substrate fit into each other structurally like the two pieces of a jigsaw puzzle or the lock and key. Just as only a particular key can fit into a particular lock because of the alignment of levers, enzymes have specific active sites, to the contours of which, the substrate molecules can fit in.

No other molecule can fit in to this slot. When a reaction mixture consists of several substrates, like A,B,C,D etc., only A and B can fit into the active sites and undergo conformational change resulting in the product P (This is an example molecule, into two products, only one active site will be present on the enzyme and only one substrate A or B can fit into the active site.)

The lock and key mechanism assumes a rigid structural configuration for the enzyme molecule and does not allow even minor adjustments between the enzyme and the substrate.

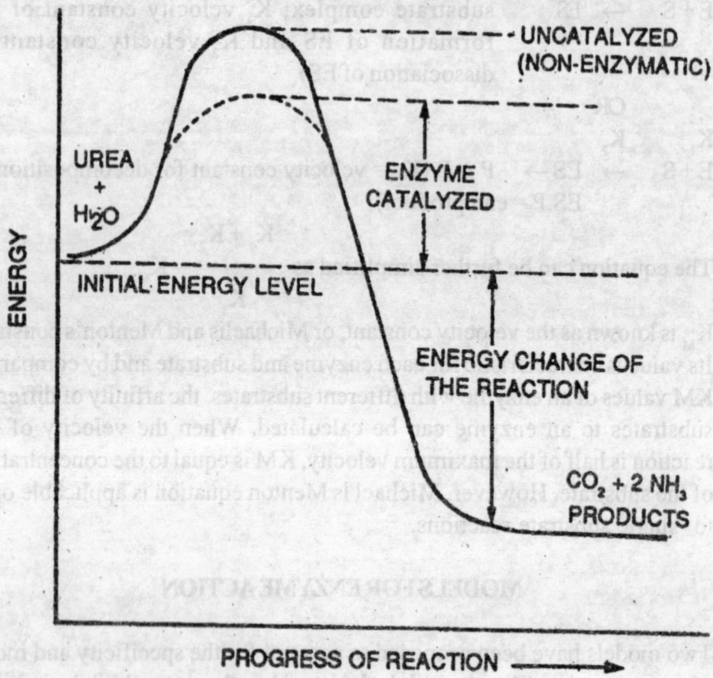

Fig. 4.4 Enzymes
Lock and Key mechanism of enzyme action

(ii) Induced fit model

Many experimental findings questioned the rigid unchangeable configuration for the enzyme molecule as proposed by the lock and key model. It has been noticed that the enzyme molecule does allow for minor adjustments in its geometry to permit the substrate to fit into the active site. Secondly, it has also been observed that the substrates induce proper orientation of the catalytic groups in the enzyme in order to activate enzyme action.

In order to make provisions for the above findings, Koshland (1959), proposed a theory called the induced fit theory. According to the induced fit model a substrate after linking itself with the enzyme alters the geometry of the enzyme protein in such a way as to achieve perfect alignment

Allosteric regulation

In reactions which involve several sequential steps many enzymes participate, each one catalyzing a particular step. Such enzymes constitute the multienzyme complexes. In multienzyme complexes, the product of one step is a substrate for the next step. The intermediate products are transferred from one enzyme to the other. In these enzyme complexes, there is end product inhibition of all the enzymes involved. In otherwords, the end product of the pathway blocks the functioning of the first enzyme (of the complex), by binding itself to a site called the *allosteric site,* which is different from the active site. The end product which binds to the allosteric site is called the *modulator* and the enzyme itself is known as the allosteric enzyme (allosterism means another space). The name allosteric enzyme was first proposed by J. Monod, J.P. Changeux and F. Jacob. Allosteric regulation is different from non competetive inhibition in that, it is the *end product* of the pathway which blocks the enzyme action of the first in the series of the enzyme complex. Allosteric regulation need not always be inhibition of enzyme action *(negative modulation).* In some cases, the end product may even activate the first enzyme of the series *(positive modulation).*

Mond, Wyman and Changeux (1965), proposed a theoritical model to explain the allosteric regulation. According to them, the allosteric enzymes are oligomers made up of identical units called monomers. The monomers are linked together in such a way that all of them occupy equivalent positions. Each monomer possesses one site for the binding of each molecule of either the substrate or the allosteric effector which are able to bind with the enzyme. The three dimensional configuration of each monomer is dictated by its association with other monomers. The oligomer molecule can exist in atleast two states which differ in the distribution and energy of interpromoter bond. When a transition from one state to another state occurs, there is a change in the affinity of the binding sites towards the molecule.

Double displacement (ping pong) reactions : In enzyme reactions involving a single substrate, it is well known that the substrate concentration influences the rate of the reactions. In many biological reactions however, a single reaction with the enzyme involves two or more substrates participating simultaneously. For example in the following reaction involving *hexokinase,* ATP and glucose are substrates with ADP and glucose - 6 - phosphate being the products.

$$\text{Glucose} + \text{ATP} \xrightarrow{\text{hexokinase}} \text{Glucose 6 phosphate} + \text{ADP}$$

In a situation as mentioned above, what will be the influence of the substrate concentration on the enzyme activity? In several bisubstrate reactions, it has been observed that the enzyme has a separate velocity constant (KM) for each of the substrates involved.

Bisubstrate reactions usually involve the transfer of a functional atom or group from one substrate to the other. This may proceed in one of the two following ways – (a) *Single displacement method* and (b) *Double displacement method.* In the first type, both the substrates bind to the enzyme simultaneously to form the enzyme substrate complex.

A + B + E EAB

(A - substrate EAB - Enzyme substrate complex B- substrate E - enzyme)

EAB then breaks up to form the products C and D

EAB —— C + D + E

In the secondary category of bisubstrate reaction called the *double displacement type,* only one substrate can combine with the active site of the enzyme at any given time. After the substrate transfers its functional group to the enzyme molecule, it (substrate) leaves the enzyme as the product. The second substrate now binds to the enzyme (which has the functional group of first substrate), accepts the functional group donated by the first substrate and leaves the enzyme as product.

E + A EA

EA E' + C

E + B E'B

E'B E + D

(A and B = substrates, C and D = products, E = enzyme EA = enzyme substrate complex, E'=enzyme with the functional group of substrate A, E'B = enzyme substrate complex; enzyme has the functional group of A). The enzyme which has the functional group of one substrate after it (substrate) has been released as the product, may be called a *partial enzyme substrate complex.*

This type of a reaction in which the substrates alternately bind and get released to the enzyme is called **ping pong reaction** or **double displacement reaction.**

Transamination reactions involving the enzyme *transaminase* provide a typical example of ping pong reactions. In a reaction involving glutamic acid and oxaloacetic acid, the enzyme glutamic transaminase has pyriodoxal phosphate as the prosthetic group. In the first of the reaction, glutamic acid binds to the prosthetic group of the enzyme (pyridoxal phosphate), and donates the amino group and is released as a α keto glutaric acid (product). In the second step, oxaloacetic acid binds to the pyridoxal phosphate (which has the amino group bound to it), accepts the amino group and is released as Aspartic acid (product).

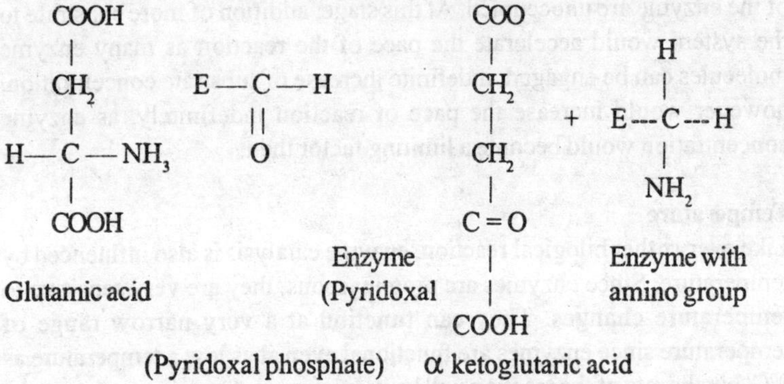

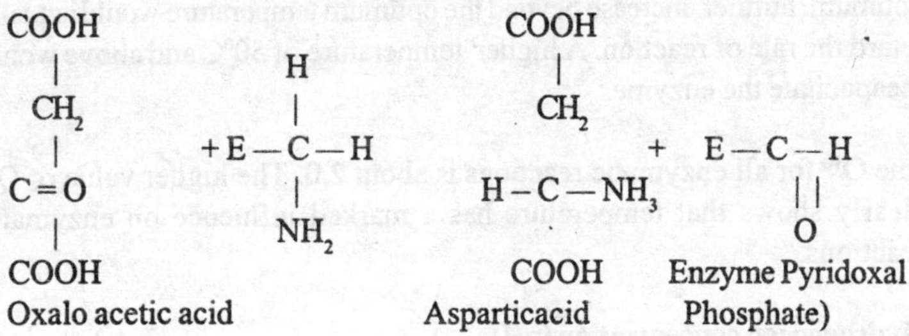

FACTORS AFFECTING ENZYME ACTION

The following factors influence the enzyme action.

1. Enzyme concentration

At specific concentrations, and in the presence of large number of substrate molecules, the reaction will proceed at a particular speed. Any addition of substrate at this stage is not of any consequence as all the enzyme active sites are already occupied. In such a situation, increasing the concentration of the enzyme will increase the rate of the reaction. This increase stops at a point when substrate molecules become a limiting factor.

Substrate concentration

As the enzyme molecules are very large in size when compared with the substrate molecules, relatively large amount of substrate is required to engage all the enzyme molecules. At a particular concentration of the enzyme, if the substrate molecules are at a low concentration, the rate of the reaction will not proceed at the required pace since many active sites of the enzyme are unoccupied. At this stage, addition of more substrate to the system would accelerate the pace of the reaction as many enzyme molecules can be engaged, indefinite increase of substrate concentration, however would increase the pace of reaction indefinitely, as enzyme concentration would become a limiting factor then.

Temperature

Like every other bilogical reaction, enzyme catalysis is also influenced by temperature. Since enzymes are proteinaceous, they are very sensitive to temperature changes. They can function at a very narrow range of temperature since enzymes are functional even at as low a temperature as 0°C, but the rate of the reaction will be its minimum. A rise in temperature at this stage would bring in a geometric progress in the rate of the reaction. Very soon, this reaches a peak and this temperature level is known as the optimum. Further, increase beyond the optimum temperature would actually retard the rate of reaction. A higher temperature of 50°C and above would incapacitate the enzyme.

The Q^{10} for all enzymatic reactions is about 2.0. The higher value of Q^{10} clearly shows that temperature has a marked influence on enzymatic reactions.

Hydrogen ion concentration (pH)

Most of the enzymes are catalytically active only at a particular pH. Sometimes, they may have a range of pH at which their activity will be sustained. Any increase or decrease from this pH would render the enzymatic activity impossible. The different enzymes, however, would vary in their pH requirement. While some enzymes are active at a low pH, others prefer an alkaline pH; still others require a neutral pH for their optimum activity.

Inhibitors

The presence of certain inhibitor molecules in the system would render the enzyme inactive. Various substances are known to inhibit enzyme action. These are of 3 kinds.

1. Competitive inhibitors
2. Non - competitive inhibitors
3. End product accumulation

In many cases, if the end products of a reaction are not removed from the system they accumulate and bring about inhibition of enzyme action known as feed back inhibition. The feed back inhibition many a time may be due to allosteric effect as can be seen in reactions involving multienzyme complexes. The end product inhibition of enzyme action may also be due to a difference in the pH value of the substrate (at which enzyme is active), and of the end product.

5

PHOTOSYNTHESIS

Ever since its origin, sometime in the remote past, life has been able to maintain itself in myriads of forms mainly because of its capacity of reproduction. An organism, however, has to exist,maintain itself in the healthy condition and only then can it reproduce. The existence of all forms of life on this planet has been made possible mainly by means of energy conversions. Organisms trap energy from environment which they utilize for their activities, and after a series of interconversions it (energy) is returned to the environment.

The phenomenon of energy input from the abiological environment to the biological system is always through the plants – green plants. Green plants thus, can be regarded as the basic fulcrum on which revolve all forms of life. In fact it is impossible to imagine what course the evolution of life would have taken but for the green plants. All forms of life are dependent on green plants (directly or indirectly) for their sustenance.

This unique phenomenon of energy input into the biological system is known as photosynthesis. The energy input has to be mediated only through the green plants, for they alone are capable of absorbing the energy from solar radiation.

Source of energy

There are only two sources of energy on this planet. One is, atomic energy and the other is solar energy. Of these two–atomic energy is not easily negotiable and available. Solar energy, however, is easily available in large quantities and living organisms have developed in the course of evolution mechanisms to trap and utilize this unique source of energy.

All forms of life, however, cannot directly harness solar energy. The capacity to trap solar energy is unique only for green plants and some bacteria. Green plants possess the green pigment viz., chlorophyll which can capture, transform, translocate and store energy which will be readily available for all activities of life.

Photosynthesis may thus be explained as the process in which the light energy will be converted into chemical energy. Except for green plants and some bacteria, all other organisms cannot directly utilize the solar energy; hence they are dependent on green plants for their nutritional requirements. Green plants and some bacteria are called autotrophic (they can prepare their own food), while all other organisms are called heterotrophic (they cannot prepare their own food). There is however one group of exceptional organisms which are autotrophic in spite of being non-green. These are the Chemosynthetic bacteria, which obtain energy by the oxidation of inorganic substrates such as ferrous ions and sulphur dissolved from the earth's crust of H_2S released from volcanic action. But considering the overall energy budget of lifeforms, the energy produced by chemosynthesis is very meagre and is of least significance quantitatively.

Definition

Photosynthesis (Photo = Light), may be defined as the process in which CO_2 and H_2O are put together (by green plants) into an organic molecule (sugar) utilizing the solar energy.

$$CO_2 + H_2O \xrightarrow[\textit{Green Plants}]{\textit{Light}} \text{Sugar (Carbohydrates)}$$

Photosynthesis may also be defined as the reduction of CO_2 to glucose ($C_2H_{12}O_6$) with the **H** necessary for reduction coming from water.

Photosynthesis involves both oxidation as well as reduction ultimately in synthesize energy rich compounds. Water is oxidized (electrons are removed), and the **H** released from water is used to reduce CO_2 to carbohydrates. Oxidation of water and the units of light energy (absorbed) produce energy intermediary compounds called ATP (Adenosine Triphosphate), and reduced coenzymes called NADPH + H_{+} (Nicotinamide adenosine diphosphate). These two compounds help in the fixation (reduction) of carbon dioxide into carbohydrates. These carbohydrates in turn constitute the starting point for the synthesis of various other metabolities necessary for the survival of the organism.

The definition and brief explanation of photosynthesis given above sums up the significance of the process. A few other definitions (different ony in the usage of the words) of photosynthesis are given below.

1. "Biosynthesis of simple carbohydrates from CO_2 and H_2O in the presence of sunight inside chlorophyll containing cells".
2. "Production of carbon containing compounds from CO_2 and hydrogen down by illuminated green cells".
3. "Photosynthesis is a redox process"
4. "Photosynthesis is a sensitized photochemical oxidation and reduction mechanism between water (H_2 donor) and CO_2 taking place at biological temperature".
5. "Photosynthesis is the formation of carbon containing compounds from CO_2 and water by illuminated green cells, water and oxygen being the by products".
6. Photosynthesis deals with the capturing and transforming of light energy into chemical energy".

The overall reaction of photosynthesis may be represented as follows

$$6CO_2 + 6H_2O \xrightarrow[\textit{Chlorophyll}]{\textit{Light}} C_6H_{12}O_6 + 6O_2$$

Oxygen which is released as a byproduct originates from water (Ruben and Kamen, 1941). In order to account for this, the equation should be changed as follows –

$$6CO_2 + 12H_2O \xrightarrow[\textit{Chlorophyll}]{\textit{Light}} C_6H_{12}O_6 + 6O_2 + 6H_2O$$

MAGNITUDE OF PHOTOSYNTHESIS

As has already been pointed out, photosynthesis is the most important biological process that supports all forms of life. Either directly or indirectly, all forms consume photosynthetic energy. From the point of view of human welfare, the contribution of photosynthesis is immense. Coal, gas oil etc., represent the photosynthetic products of the plants belonging to early geological periods.

The significanse of photosynthesis as a vital process contributing to the survival of life can be understood when we take into consideration the overall; rate of photosynthesis.

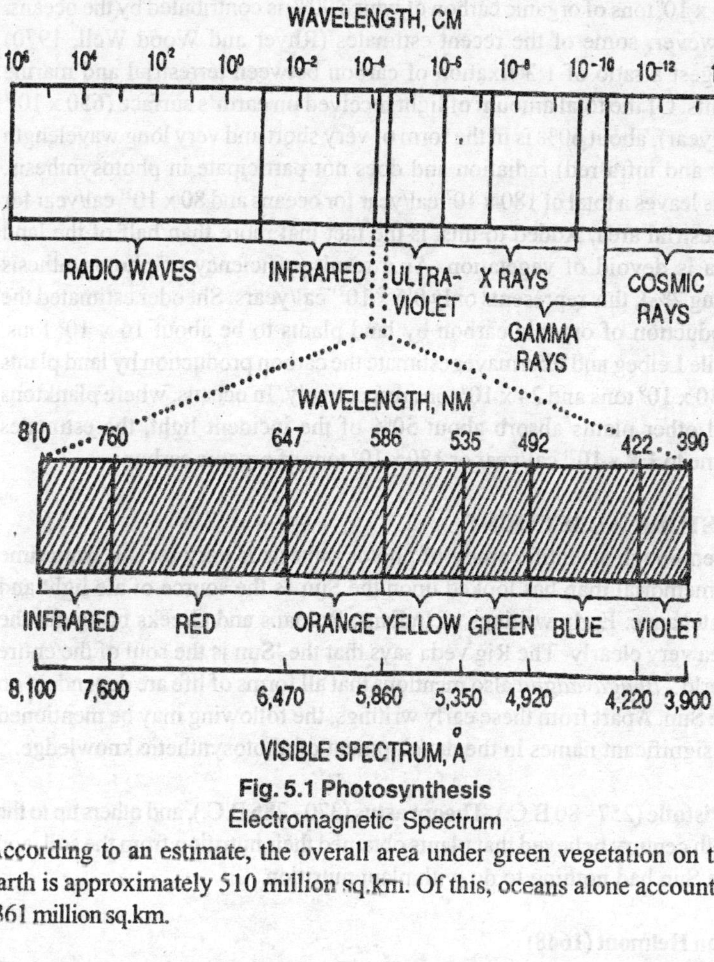

Fig. 5.1 Photosynthesis
Electromagnetic Spectrum

According to an estimate, the overall area under green vegetation on this earth is approximately 510 million sq.km. Of this, oceans alone account fo r361 million sq.km.

Regarding the availability of light, solar radiation which is emitted from the sun in the form of electromagnetic radiation reaches the outer atmosphere. In terms of energy, the outer atmosphere receives about 1300 x 10^{21} cal/year. Much of this is shielded by the atmosphere and the surface of the earth receives light equivalent to 650 x 10^{21} cal/year. Considering the distribution of water and land masses, marine plants receive about 90 x 10^{21} cal/year and terrestrial green vegetation receives about 25 x 10^{21} cal/year. If we assume photosynthetic efficiency is 2% of this energy, it amounts to 50 x 10^9 tons of organic carbon for terrestrial green plants and 180 x 10^9 tons for marine plants.
Rabinowitch (1951), calculates that photosynthesis produces approximately

155 x 10^9 tons of organic carbon of which 90% is contributed by the oceans. However, some of the recent estimates (Rhyer and Wood Well, 1970) suggest a ratio of 1:3 fixation of carbon between terrestrial and marine plants. Of the total amount of light received on earth's surface (650 x 10^{21} cal/year), about 60% is in the form of very short and very long wavelength (far and infra red) radiation and does not participate in photosynthesis. This leaves a total of 180 x 10^{21} cal/year for oceans and 80 x 10^{21} cal/year for terrestrial area. Added to this, is the fact that more than half of the land area is devoid of vegetation. As a result (efficiency of photosynthesis being 2%), this represents only 0.5 x 10^{21} cal/years. Sheoder estimated the production of organic carbon by land plants to be about 16 x 10^9 tons, while Leibeg and Ebermayer estimate the carbon production by land plants at 30 x 10^9 tons and 24 x 10^9 tons respectively. In oceans, where planktons and other plants absorb about 50% of the incident light, the estimates come to 1.8 x 10^{21} cal/year or 180 x 10^9 tons of organic carbon.

HISTORICAL ACCOUNT

Eventhough scientific study of photosynthesis is very recent, from time immemorial man has looked upon the Sun as the source of ale light and sustenance. Early writings of Indians, Romans and Greeks bring out the idea very clearly. The Rig veda says that the 'Sun is the soul of the entire world'. *Bhagavadgita* also mentions that all forms of life are dependent on the Sun. Apart from these early writings, the following may be mentioned as significant names in the development of photosynthetic knowledge.

Aristotle (257 - 80 B.C.), **Theoprastus** (370 - 285 B.C.), and others up to the 17th century believed that plants obtained their nutrition from the soil, and the Sun had nothing to do with plant nutrition.

Van Helmont (1648)

In this experiments with willow seedlings, he observed that water, and not the soil was the principle source of nutrition.

Wood Ward (1699)

Opined that the plant body is made up not of water, but a peculiar soil substance which is absorbed along with water.

Stephen Hales (1927)

Who is regardes as the **Father of Plant Physiology** proposed that green plants obtained their nutrition partly from the air through the leaves, and sunlight may have a role in it. He, however, was not aware of the part played by carbon dioxide.

Joseph Priestley (1972)

Who conducted several experiments on mint plants stated that plants can purify air. He also observed that impure air (CO_2) would get purified (vital air - O_2) if kept in contact with the green mint plants. Priestley also knew nothing about the role of either CO_2 or light in photosynthesis.

Jan Ingen Housz (1771)

A Dutch physician, demonstrated that plants purified air only in the presence of light. He also stated that plants too contributed to 'bad air' during darkness.

Jean Zenebier (1782)

A Swiss minister agreed with the findings of Ingen Housz, and commented that plants absorb CO_2 from the atmosphere during daytime and exhale it (CO_2) during darkness. The role of two gases (O_2 and CO_2) had been implicated in plant nutrition by 1782. Works of Lavoisier and others identified that these gases were O_2 and CO_2.

N.T. de Saussure (1804)

Identified that water is an active participant in photosynthesis together with CO_2. He also made the first quantitative measurements of photosynthesis and opined that approximately equal volumes of CO_2 and O_2 are exchanged during photosynthesis.

Mayer (1842)

Who formulated the law of conservation of energy, observed that during photosynthesis solar energy is stored in the form of chemical energy.

In 1871, two French Chemists, Joseph Bienaime Caventou, and Piere Pelletier isolated the chlorophyll pigment from green plants. Richard Willstatler obtained chlorophyll in the pure form. He also identified two forms – **Chlorophyll a** and **Chlorophyll b** which differed in the manner of their light absorption.

Leibeg (1843)

Stated that the source of all carbon in plants is the CO_2 of the atmosphere.

Julius Sachs (1846)

A German plant physiologist, observed the growth of starch grains in choloroplasts, only in the area of the leaf exposed to light. Thus, he was the first one to note the increase in organic content due to photosynthesis.

Boussingault (1846)
Calculated the ratio of CO_2 intake and O_2 release, and came to the conclusion that it is 1:1.

Baeyer (1870)
Proposed the formaldehyde hypothesis. He argued that CO_2 gets reduced and forms CH_2O which on polymerization forms the glucose molecule. The formaldehyde hypothesis, however, has been abandoned not only because it has not been detected in plants, but it is also poisonous (to the plants), if it is present.

Blackmann (1905)
Identified that photosynthesis involves two distinct phases if it is present.

Blackmann (1905)
Identified that photosynthesis involves two distinct phases *viz.*, light phase and dark phase.

Van Niel (1931)
Showed that photosynthesis is a biological oxidation reduction reaction.

Robin Hill (1937)
Showed that photosynthesis is localized in the chloroplasts. In the presence of certain electron acceptors (oxidizing - reducing agents), isolated chloroplasts accept light energy and release O_2. This phenomenon of illuminated (isolated) chloroplasts splitting water and releasing O_2, has come to be known as Hill's reaction.

Samuel Ruben, Randall Kamen and Hyde (1941)
Showed (using radioisotopes of oxygen) that the oxygen released during photosynthesis comes only from water and not from CO_2.

Consden, Gordon and Martin (1944)
Developed paper chromatography technique for the separation of amino acids. This technique later was employed to detect the various intermediate compounds during the dark fixation of CO_2.

Arnon (1951 - 60)
Developed the concept of photophosphorylation and showed that coenzyme II picks up the H released by splitting up of water.

Benson, Bassham, Melvin Calvin and others have worked out in detail the carbon reduction cycle with the help of techniques like chromatography and radioautography.

RAW MATERIALS FOR PHOTOSYNTHESIS

There are some basic requirements (raw materials) for the process of photosynthesis. These are CO_2, a hydrogen donor (water), and light energy, besides of course the structural framework of the green plant in the form of chloroplasts.

Carbondioxide

The only source of carbon for plants for the purpose of assimilation is the atmospheric carbon dioxide. Approximately about 20×20^{11} tons of carbon dioxide is used for the production of 50×10^{9} tons of organic carbon. Besides atmosphere, other sources of carbon are biological respiration decaying organic matter etc. Carbon dioxide is in the gaseous form in the atmosphere, while it is in dissolved condition in water.

Carbon dioxide is present in trace quantities in the atmosphere (o.0-35%). There has been a gradual increase in the concentration of the atmospheric carbon dioxide since the middle of 19th century perhaps due to increased pollution (Wood Well, 1970).

Hydrogen Donor

In almost all photosynthetic organisms, water is the hydrogen donor. Water splits into H and OH ions in the presence of light. H is picked up by the enzyme nadph for reduction of CO_2, while OH recombines to form water and oxygen.

Certain chemosynthetic bacteria, however, use compounds of Iron and Sulphur as hydrogen donors.

Light

Solar radiation is the only source of light for all forms of life. The radiation from Sun is emitted in the form of electromagnetic radiation with radiowaves at one end and cosmic rays at the other. Much of this radiation light shielded by the outer reaches of the atmosphere and only visible Light may be defined as the visible (to human eye) part of the electromagnetic spectrum. The visible spectrum is confined between wavelengths of 390mμ and 760 mμ. Spectral analysis of the visible light shows the following rays with their respective wavelengths given in parenthesis.

Violet (390 - 430 mμ)
Blue (430 - 470 mμ)
Bluegreen (470 - 500 mμ)
Green (500 - 560 mμ)
Yellow (560 - 600 mμ)
Orange (600 - 650 mμ)
Red (650 - 760 mμ)

Rays of the radiation with wavelengths shorter than violet are called U.V. (Ultra violet - 100 A⁰ - 390 A⁰), X rays (0.1 A⁰ - 100 A⁰), Gamma rays (0.001 A⁰ - 0.1 A⁰) and cosmic rays (less than 0.001 A⁰). Radiation with wave lengths longer than red are called infra red (7600 A⁰ - 1000,000 A⁰). Beyond this, are electric and radio waves, which are measured (in terms of wavelength) in kilometers.

Radiation travels in the form of discrete corpuscles or particles called **Quanta or Photons.** Each quantum represents a unit of energy. Energy quantity of each ray depends on its wavelength; shorter the wavelength, higher is the energy. The wavelengths of radiation are so small that they are measured not in terms of millimeters or centimeters, but in terms of microns (μ), millimicrons (mμ) or nanometers (nm) or Angstrom (A⁰) units.

$1\ A^0 = 10^{-10}$ or 10^{-8} cms.

Each photon has an energy constant $E = hc/\lambda$, where h is planck's constant ($=6.625 \times 10^{-27}$ erg/sec) c is velocity of light (3×10^{10} cms/sec), and λ is the wavelength as given by Einstein.

The energy value of different light rays is given in the following table.

Energy Value of Different Wavelengths of Light

Sl.No.	Wavelength in mm	Approximately colour	Energy in cal
1.	300	Ultra Violet	95,200
2.	420	Violet	68,000
3.	470	Blue	60,760
4.	530	Green	53,890
5	620	Orange	46,060
6.	700	Red	40,800
7.	683	Red	41,810
8.	672	Red	42,490
9.	650	Red	43,940
10.	850	Infra Red	33,590
11.	1000	Infra Red	28,560

THE PHOTOSYNTHETIC APPARATUS

Within the cell of the green plants are present certain unique organelles called plastids. These perform many vital functions in cell physiology. Among the plastids, the most important from the point of view of photosynthesis are chloroplasts. All the physiological reactions, starting from the absorption of light energy to the fixation of CO_2 take place in the chloroplasts. No wonder then, chloroplast is aptly called the photosynthetic apparatus.

CHLOROPLAST

These are by far, the commonest and the most important among plastids. As the primary sites for trapping and converting solar energy, they are very vital for the existence of not only the green plants, but for the whole living world.

Shape

Chloroplasts have varied shapes. The algal chloroplasts exhibit an array of shapes. They may be cup shaped (*Chlamydomonas*), girdle shaped

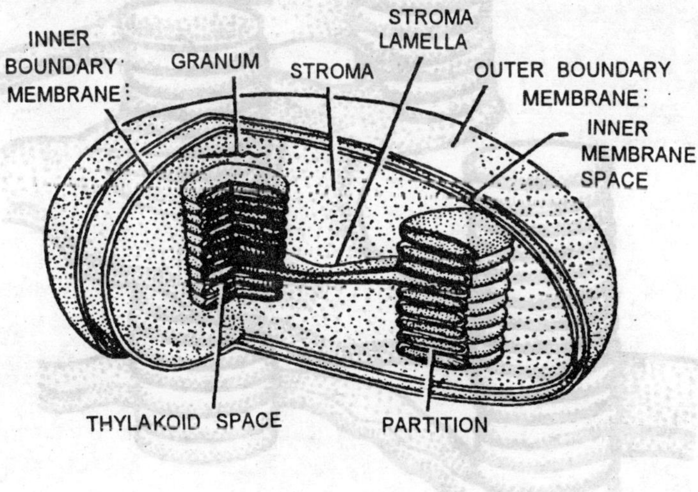

Fig. 5.2 Photosynthesis
Model of Ultrastructure of chloroplasts

Size

There is a great variety in size. Normally, they are about 1 μm thick and 4–6μm in length. Chloroplasts of polyploid cells are generally larger than in the diploid cells. Cells in the shaded area have longer chloroplasts but with less intensity in colouration.

Distribution

They are uniformly, distributed all over cytoplasm, but in some instances they cluster towards the nucleus. The concentration of chloroplast will also depend on light intensity.

Number

The number of chloroplast per cell varies. In some algae, like *Chlamydomonas,* there is only one chloroplast per cell, or two as in

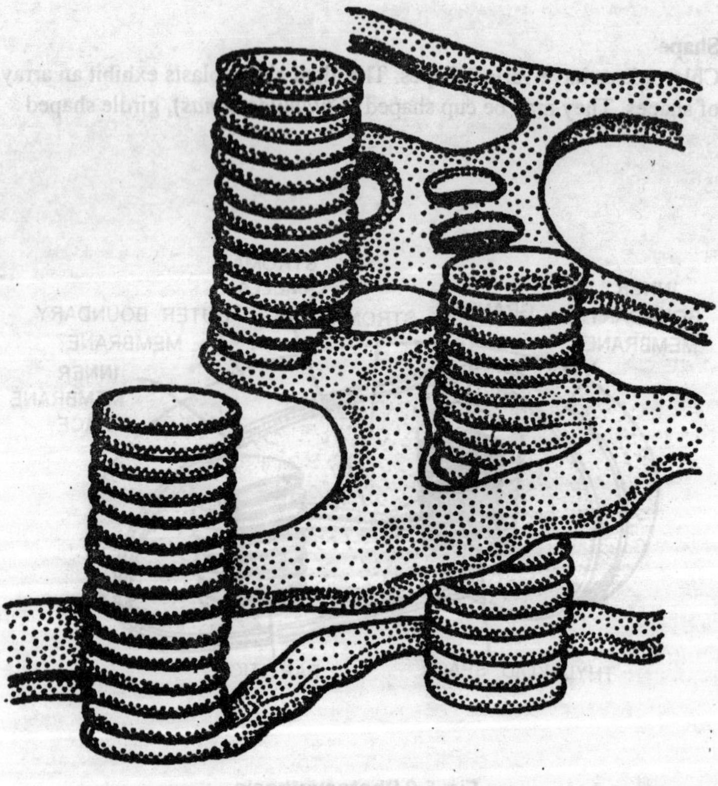

Fig. 5.3 Photosynthesis
Three dimensional view of Grana (enlarged view)

Zygnema. In higher plants, cells have a chloroplast. The leaf of *Ricinus communis* has nearly four lakh chloroplasts per square millimetre area.

Chemical composition

The chloroplasts contain carbohydrates, lipids, proteins, chlorophyll, Carotenoid and xanthophylls. The protein lipid ratio is 40:30. Besides the above, nuclei acids have also been demonstrated in the chloroplasts.

Ultra structre of the chloroplasts

The chloroplast has a covering of two membranes with an inner membrane space. These membranes are smooth and there are no perforations or particles. The membranes are differentially permeable (Mudrak 1969).

A section of the chloroplast reveals an intricate system of membranes enclosed in a granular matrix. These membranes are called *lamellae* and the surrounding matrix – the *stroma.* In a sectional view, the lamellae can be seen packed and these stacks are called *thylakoids.* In the chloroplasts of higher plants, the thylakoids themselves form highly compact bundles called **grans.** Some thylakoids of granum extend into the stroma and maintain contact with other grana. These are called – Stroma thylakoids or stroma lamellae or intergrana.

The chloroplasts of algae lack the granum arrangement and in cyanobacteria (blue green algae), the thylakoids lie naked in cytoplasm without any envelope. In such instances, pigments are uniformly distributed on or in the lamellae.

The stroma has in addition to the lamellae – granules (globuli), lipid droplets, starch grains and vesicles.

Structure of the lamellae

All the photosynthetic pigments are concentrated in the chloroplast, lamellae as well as the outer membranes are composed of lipoprotein, sub units. Each sub unit according to Weir and Benson, has a protein core surrounded by a lipid sheath. The chloroplast membranes have a single layer of sub units, whereas in thylakoids, they are double due to the appressing of two membranes. The appressed areas are called partitions. These are hydrophobic in nature. The space between the two membranes of a thylakoid (fret membrane), enclose a space called fret channel. The space between two thylakoids is called loculus while the end of disc shaped thylakoids are called margins. These areas (loculus, fret channel) are hydrophilic. The chlorophyll molecules are concentrated in the fret cannels i.e., in the space between the two membranes of thylakoids. The chlorophyll molecules may be entirely in space, or the heads may be in the fret channel, while the tails are buried into the sub units. Occassionally, the entire chlorophyll molecule may be found in the fret membrane. Other pigments like cytochromes, carotenoids etc are also found in the fret membranes.

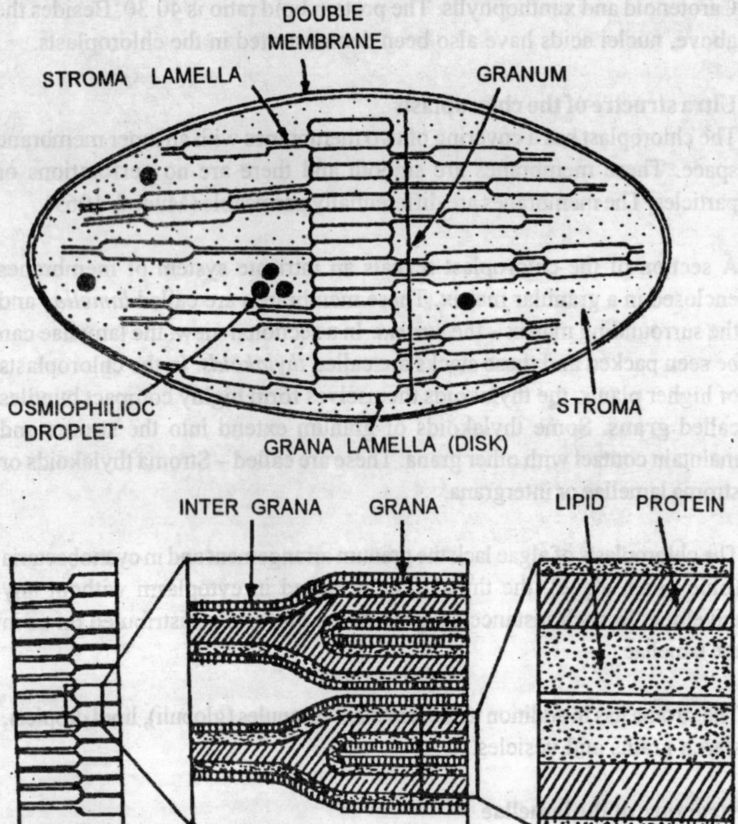

Fig. 5.4 Photosynthesis

Diagrammatic representation of Ultrastructure of chloroplast showing the lamella,

A. Entire chloroplast, B-D. Show successive enlargement of a part of the granum

Four sub units of the partition consitute a photosynthetic unit or a *quantosome*.

Pigments in chloroplast

Some of the important pigments present in chloroplasts are chlorophylls, Carotenoids, cytochromes etc. In algal cells, the thylakoids also have phycobilins.

Chlorophylls

These are the most important pigments involved in the basic photochemical reaction that sustains life. Atleast nine types of chlorophylls have been distinguished so far – chlorophylls *a,b,c,d* and *e.,* bacterio chlorophylls *a* and *b,* and the chlrobium chlorophylls 650 and 660 μm.

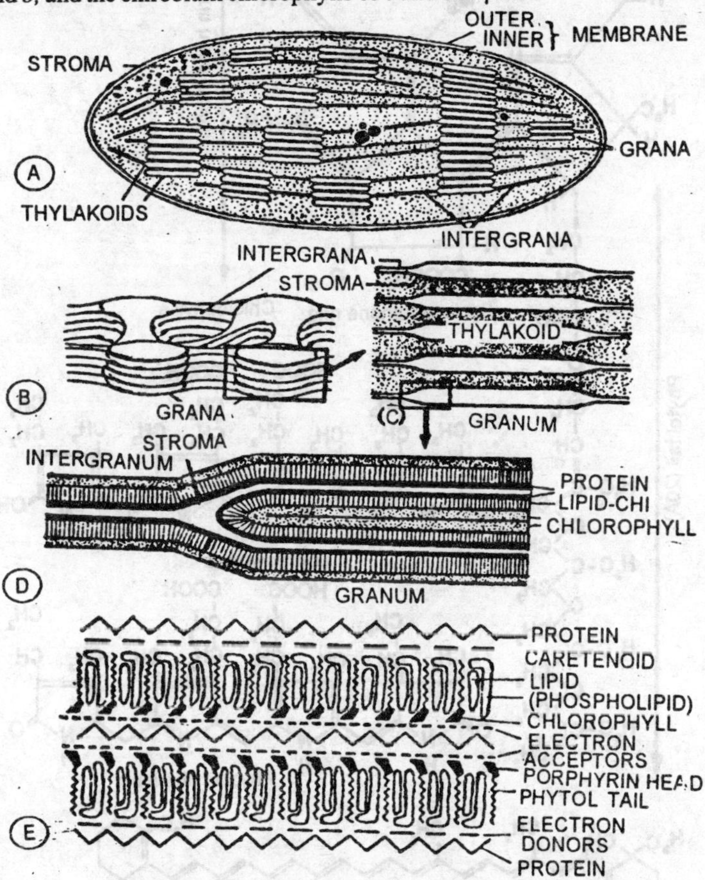

Fig. 5.5 Photosynthesis

Stages in the development of chloroplast

A. Entire chloroplast, **B.** Grana and Intergrana (diagrammatic), **C.** Section through a granum, **D.** Lamellar structure (in detail), **E.** Model for the arrangement of pigments in a granum

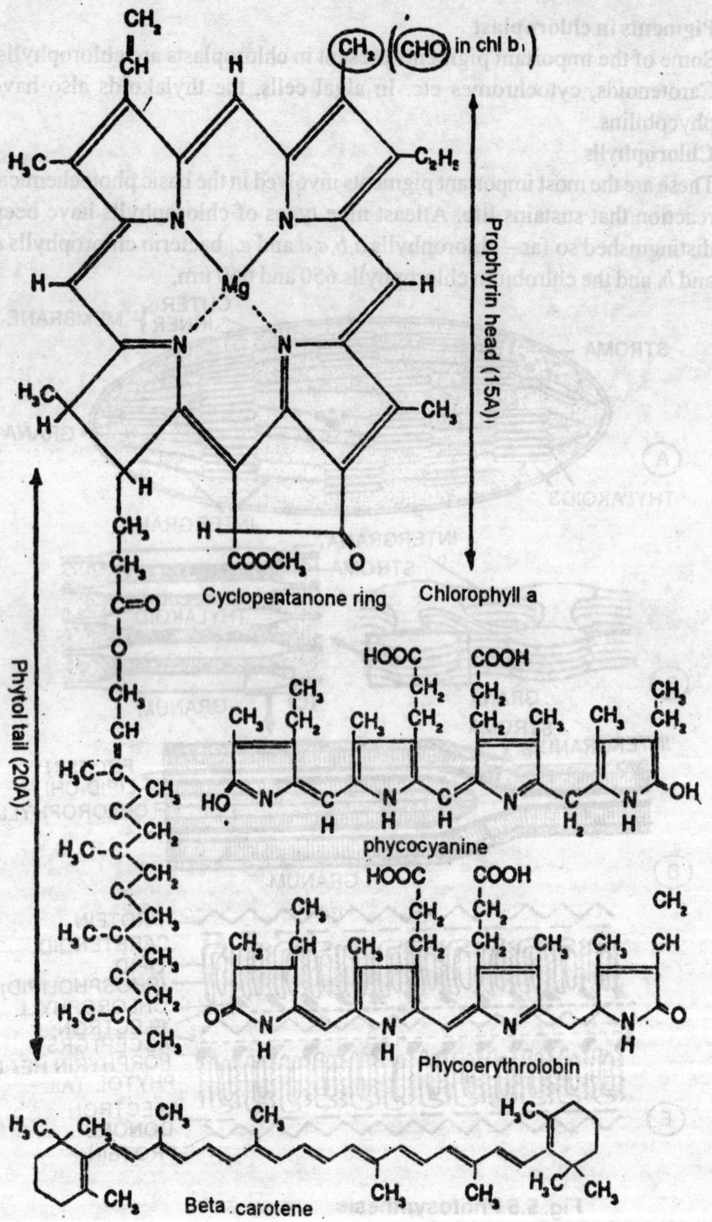

Fig. 5.6 Photosynthesis
Molecular structure of photosynthetic pigments

Among the chlorophylls, chlorophylls, and b are best known and most widely distributed. These are absent in pigmented bacteria. Chlrophylls *c*, *d* and *e* are found only in algae in combination with chlorophyll a. Bacterio chlorophylls and chlorobium chlorophylls are found only among photosynthetic bacteria.

The chlorophyll molecule is a complex structure basically made up of a 'head' and a 'tail' resembling a tennis racquet. The head is a porphyrin structure made up of four Pyrrole rings attached to each other at the centre by an isocyclic ring containing Mg atom at the centre. Extending from one of the pyrrole rings is the tail–the alcoholic chain (Phytol). The emperical formula for the chlorophyll molecule is $C_{55}H_{72}O_5N_4$ Mg. The phytol chain is esterified on the C atom of one of the pyrrole rings and has only one double bond.

Chlorophyll *a* and *b* differ structurally in having different atomic groupings at the C_3 atom in the pyrrole ring. In chlorophyll *a*, C_3 atom has a methyl group, while chlorophyll b has an alehyde group; besides the two pigments have different absorption spectra. The peaks are as follows.

Chlorophyll *a* 429, 410, and 66nm
Chlorophyll *b* 430, 453 and 442 nm

(The above spectra are of chlorophyll *in vitro*)

Carotenoid pigments

These are lipid compounds ranging in colour from yellow to purple. They are widely distributed in both plants and animals. They are also present in microorganisms including red algae, cyanobacteria, photosynthetic bacteria, fungi etc. Wackenroder (1931), isolated the first Carotenoid – carotene from, the carot tissue.

Carotenoids are derivatives of lycopene, a red pigment found in many plants. They (Carotenoids) are known to have eight isoprent (CH2=C [CH3] - CH = CH2) like residues.

The main Carotenoid found in plants is the orange yellow coloured β carotene with some quantity of α carotene.

The Carotenoids are also located in the chloroplast. According to Goodwin (1960), Carotenoids and chlorophylls may be combined with the same protein to form a complex known as photosynthein.

Carotenoids protect the chlorophyll from photooxidation and transfer the light energy they absorb to chlorophyll *a*.

Phycobilins

These are found only in algae. There are two known types of phycobilins – the phycoeroythrin (red), and phycocyanin (blue). These pigments are strongly associated with a protein and consequently it is difficult to isolated the pigment in the pure state.

Like Carotenoids, phycobilins are also involved in the transfer of light energy (they absorb) to the chlorophyll.

The absorption spectra of phycobilins presents an interesting study considering the fact that they act as accessory pigments in photosynthesis. R - phycoerythirn has peaks at 495, 540 and 545 nm, while R-phycocyanin has peaks at 550 and 615 nm.

Photosynthetic units

A photosynthetic unit may be defined as "the smallest group of coordinating pigment molecules necessary to effect a photochemical act, i.e. absorption and transportation of a light quantum to a trapping center where it causes excilation and release of an electron".

Emerson and Arnold (1932), who experimented in Chlorella plants, observed that about 2,500 molecules of the chlorophyll pigment are necessary to fix a molecule of CO_2 and they termed this number (2,500), as a photosynthetic unit. They also observed that to release one molecule of O_2, 10 quanta of light are necessary. Some of the recent works, however have shown that on the basis of absorption of 10 quanta, photosynthetic unit should consist of 230 molecules of the chlorophyll pigment. Park and Beggins (1964), working on the chloroplasts opined that photosynthetic units are distinct morphological identities and called them quantasomes. The quantasomes are 180 A^0 broad and 100 A^0 thick (`8 x 16 x 10nm). Each quantasome has 230 molecules of chlorophyll (160 chlorophyll a and 70 chlorophyll b), 48 carotenoids, 46 quinone compounds, 116 phospholipids, 144 digalactosyl diglycerides, 346 monogalactosyldiglycerides, 48 sulfolipids, unidentified lipids and many sterols. Quantasomes have a molecular weight of about two millions and are arranged in a monolayer on the thylakoid bodies. Electron microscopic studies have revealed that the individual quantasome may have four or more sub units. This middle region of the quantasome is called the **reaction centre.**

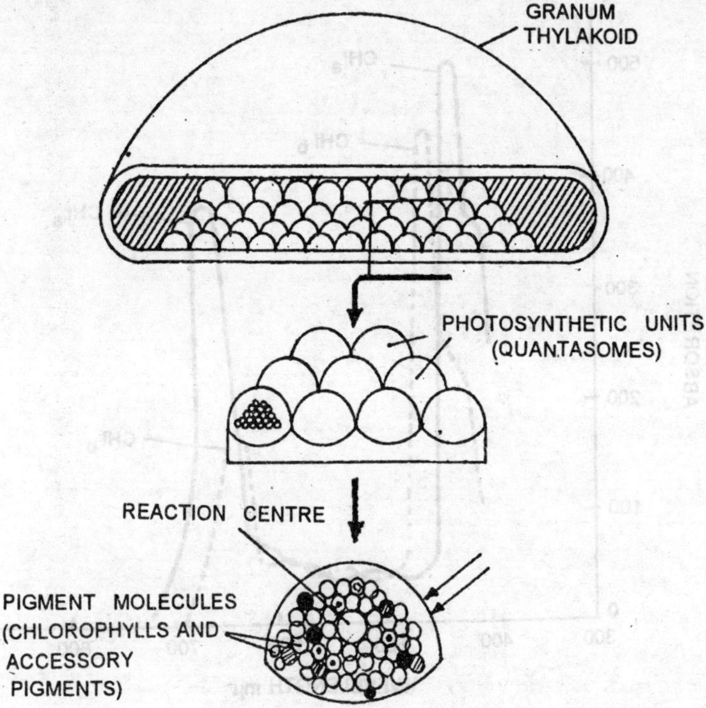

Fig. 5.7 Photosynthesis

Diagrammatic representation of a thylakoid, quantasomes and the reaction centre

Robert Huber, Johann Deisenhofer, and Jartmutt Michel were awarded the 1988 chemistry Nobel Prize for unravelling the secret of the reaction centre. Working at the Max Planck Institute in Martinsried, Germany, these scientists, made their breakthrough by extracting the reaction centre from the membranes of certain photosynthetic bacteria. They then crystallized it and analysed the structure using X - ray diffraction.

Photosynthetic pigments

The most important pigments involved in photosynthesis are chlorophylls and carotenoids. Other pigments present, but not involved in photosynthesis are biliproteins or chromatoproteins.

Chlorophylls

There are five important types of chlorophylls. These are Chlorophyll *a,b,c d* and *e*. In photosynthetic bacteria, two Chlorophyll–like pigments– bacteriochlorophyll and bacterioviridin have been reported.

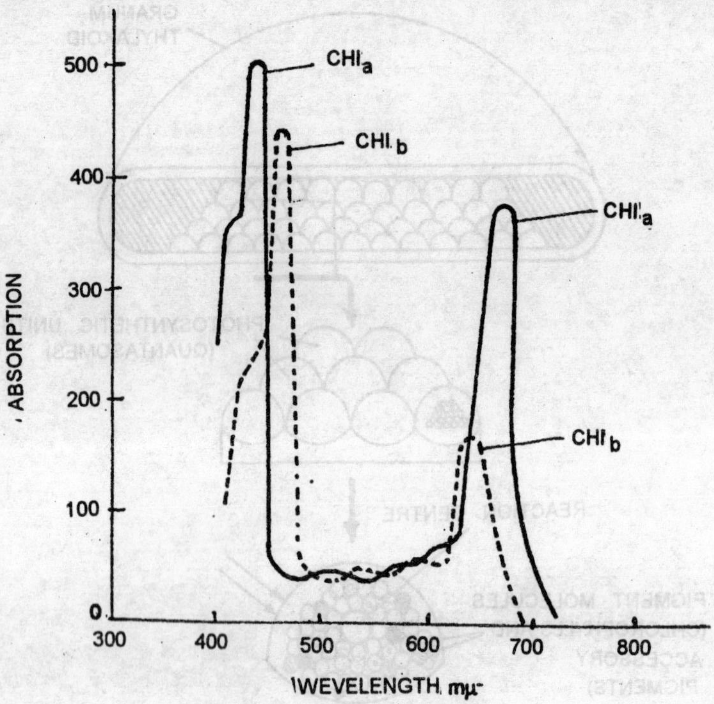

Fig. 5.8 Photosynthesis
Absorption spectra of chlorophyll a and b

Chlorophyll *a* is present both in higher plants and algae. The pigment is insoluble in water but readily soluble in either. It appears blue green in transmitted light, but reddish in reflected light. Chlorophyll *a* is the principal pigment involved in the trapping of light energy.

Chlorophyll *b* is also found in green and higher plants. It is yellowish green in transmitted light but reddish in reflected light. It is similar to chlorophyll a in solubility. Chlorophyll *b* is absent in Cyanophyceae, Rhodophyceae Phaeophyceae and diatoms.

Chlorophyll *c*, *d* and *e* are present in brown, red and golden brown (Xanthophyceae) algae respectively.

Carotenes and Xanthophylls. They are insoluble in water but soluble in organic solvents.

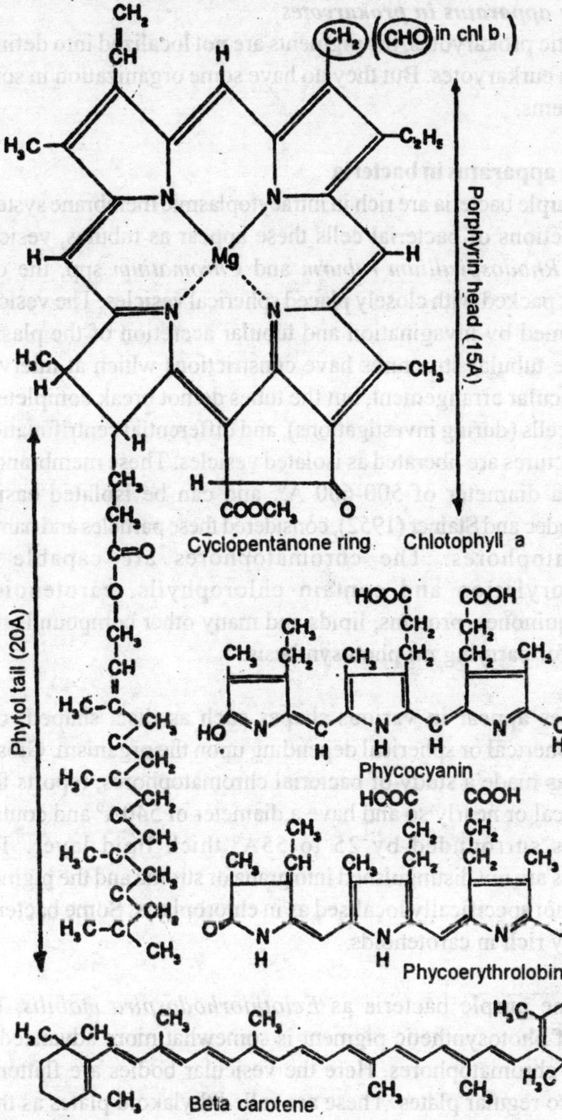

Fig. 5.9 Photosynthesis
Molecular structure of photosynthetic pigments

Photosynthetic apparatus in prokaryotes

In photosynthetic prokaryotes, the pigments are not localized into definite organelles as in eurkaryotes. But they do have some organization in some membrane systems.

Photosynthetic apparatus in bacteria

Phototrophic purple bacteria are rich in intracytoplasmic membrane system. In ultra thin sections of bacterial cells these appear as tubules, vesicles and plates. In *Rhodospirillium ruburm* and *Chromatium* spp, the cell lumen is almost packed with closely placed spherical vesicles. The vesicles seem to be formed by invagination and tubular accretion of the plasma membrane. The tubular structures have constrictions which at intervals disturb the vesicular arrangement, but the tubes do not break completely. On lysis of the cells (during investigations), and differential centrifugation, the tubular structures are liberated as isolated vesicles. These membranous vesicles have a diameter of 500-600 A^0, and can be isolated easily. Schachman, Pradec and Stainer (1952), considered these particles and named them chromatophores. The chromatophores are capable of photophosphorylation and contain chlorophylls, carotenoids, cytochromes, quinones, proteins, lipids and many other compounds that are necessary for carrying on photosynthesis.

Chromatophores appear in various shapes such as disc, shaped, cup shaped, hemispherical or spherical depending upon the organism. Gibson (1965), who has made a study of bacterial chromatophores, reports that they are spherical or nearly so and have a diameter of $540A^0$ and contain core pigments surrounded by 25 to $55A^0$ thick lipid layer. The chromatophores are not distinguished into grana or stroma and the pigment molecules are not specifically localised as in chloroplasts. Some bacterial species are very rich in carotenoids.

In some marine purple bacteria as *Eclothiorhodospira mobilis,* the organization of photosynthetic pigment is somewhat more advanced in comparison to chromatophores. Here the vesicular bodies are flattened and stacked into regular plates. These are called thylakoid plates as they bear a resemblance to thylakoids of eukaryotes. Here, 10-16 thylakoid plates are stacked into bundles resembling the granum of chloroplasts.

In the cells of green sulphur bacteria. the photosynthetic pigments are bound to two separate cell structures. The antenna pigments are concentrated in the chlorosomes, whereas the pigments of reactions centre are in ther cytoplasmic membrane.

The photosynthetic pigments of phototrophic purple bacteria are bound to intracytoplasmic membranes.

Photosynthetic apparatus in cyanobacteria

In cyanobacteria, the peripheral region of cytoplasm contains the photosynthetic apparatus. In unicellular forms, this commonly consists of a few membrane layers which extend in concentric sheets below the cell membrane and surround the central nucleoplasmic zone. In filamentous members, the cells have more elaborate membrane systems formed due to expansion and invagination of the membranes. The basic structure of the membranes looks like a flattened sac and is designated as the thylakoid. The membranes which are closely thylakoid sac.

The thylakoid lamellae have a lipid bilayer incorporating the lipophilic photosynthetic pigments – chlorophyll a and various carotenoids. In addition, it also incorporates cytochromes, plastocyanin and other components of the electron transport chain. Chlorophyll *a* which is the only chlorophyll present in cyanobacteria exists in three spectral variations (with peaks at 670, 680 and 700nm). Among the carotenoids, which absorb in the range between 480 and 520nm, β carotene appears to be widely distributed in cyanobacteria. Xanothophylls, however, differ in their abundance as well as presence in various species of cyanobacteria. Some of the xanthophylls seen in cyanobacteria are echineone, zeaxanthin, oscillaxanthin, myxoxanthophyll etc.

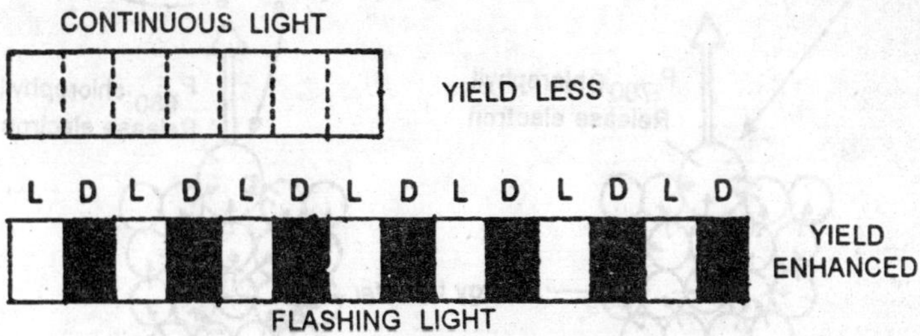

Fig. 5.10 Photosynthesis

Red drop and Emerson effect in Chlorella

A significant part of the photosynthetic pigment complement of cyanobacteria is located in some granular supermolecular complexes such as phycobilisomes. These contain the chromophore bearing water soluble phycobiliproteins namely blue phycocyanin, allophycocyanin and phyceorythrins. The relative quantities of these pigments may vary according to species and the spectral composition of light. It is that which determines the varied colour appearance of the organisms.

The table belwo gives an account of the various classes of pigments present in cyanobacteria.

The major phrotosynthetic pigments of blue - greens.

Group	Class	Major absorption bands (nm)
Chlorophylls	Chlorophylla	435, (670), 680, (700)
Carotenoids	β-carotene	(431), 450-454, 478-480
	Echinenone	455-459, (475)
	Zeaxanthin	(430), 453, 479
	Canthaxanthin	466
	Myxoxanthophyll	504-508, 474-476, 450-452
Phycobiliproteins	Allophococyanin	650
	Phycocyanin	610-625
	Phycoerythrin	555-565

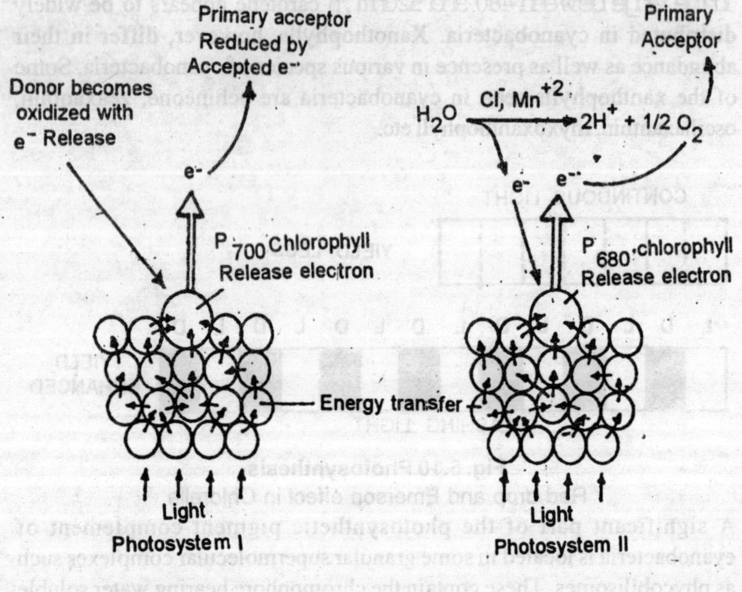

Fig. 5.11 Photosynthesis

Energy transfer in Photosystem I and Photosystem II in green plants

Phycobiliproteins represent, the primary light harvesting or antenna pigments, for photosystem II in cyanobacteria. Light energy trapped by these pigments is transferred to chlorophyll with a very efficiency. Each phycobilisome consists of a triangular core surrounded by six peripheral rods and is made up of disc shaped subunits. The triangular core which is in direct contact with the thylakoid membrane has the pigments allophycocyanin, while the rods contain phycocyanin and phycoerythrin. The phycobilisomes are arranged in parallel rows on the outer surface of the thylakoid membrane. Close proximity between phycobilisomes and thylakoid membranes is necessary for maximum energy transfer in the primary act of photosynthetic.

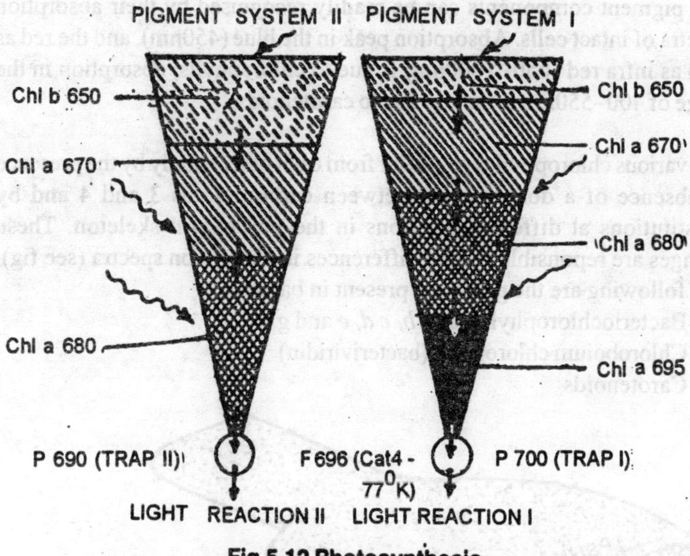

Fig.5.12 Photosynthesis
Distribution of different pigments in PS I and PS II

Complementary chromatic adaption

Environmental conditions such as mineral nutrients, and intensity and quality of light is known to affect the pigment composition in cyanobacteria. It has been suggested that the changes enable the most efficient utilization of available light energy. The synthesis of phycobilin pigments and even the formation of thylakoids is susceptible to environmental influence. Change in pigmentation is largely due to change in phycocyanin phycoerythrin ratio. Certain strains appear distinctly bluish-green when grown under red light because phycocyanin which absorbs minaly red light is present in large quantities. On the other hand, strains grown under green light will appear reddish due to the preponderance of green light absorbing phycoerythrin in light cells. Thus, the organisms produce a pigment which can absorb best the existing light. The shift between red light and green light effects was found to occur at 590nm wavelength. The phenomenon of this shift is called *complementary chromatic adaptation* or *Gauidukov phenomenon* (named after the discoveror). This phenomenon is seen in only those cyanobacteria which synthesize phycoerythrin.

Photosynthetic pigments is bacteria

Suspensions of photosynthetic bacteria appear green, purple, violet, red, brown or salmon coloured to the eye based on the pigment composition. The pigment components can be readily recognised by their absorption spectra of intact cells. Absorption peak in the blue (450nm), and the red as well as infra red (650- 1000nm) is due to chlorophylls; absorption in the range of 400–550nm is mainly due to carotenoids.

The various chlorophyll types differ from each other mainly by the presence or absence of a double bond between carbon atoms 3 and 4 and by substitutions at different positions in the poryphyrin skeleton. These changes are reponsible for the differences in absorption spectra (see fig). The following are the pigments present in bacteria.

1. Bacteriochlorophyll *a* and *b, c d, e* and *g*
2. Chloroboium chlorophyll (bacteriviridin)
3. Carotenoids

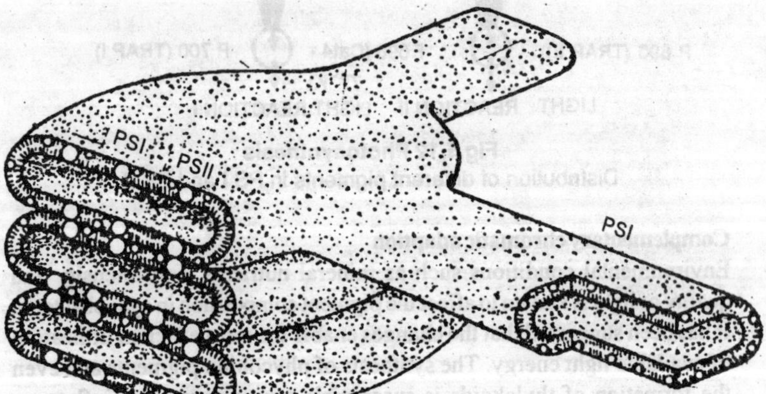

Fig. 5.13 Photosynthesis

Model of the structure of Granum thylakoid and a stroma thylakoid showing the distribution of PS I and PS II

Bacteriochlorophulls

These are found only in purple sulphur and non sulphur bacteria. The types of bacteriochlorophylls differ in their absorption spectra which is given in the following table :

Types of Bacteria Chlorophylls

Designation of Jenson et. Albugo, 1964	Former Designation	Characteristic Adsorption Maxima in living cells (nm)
Bhcl a	Bacteriochlorophyll	375, 590, 805, 830-890
Bhcl b	Bacteriochlorophyll *b*	400, 605, 835-950, 1020-1040
Bhcl c	Chlorobium chl 660	Long wavelength absorption max 745-755
Bhcl d	Chlorobium chl 650	– do – 705-740

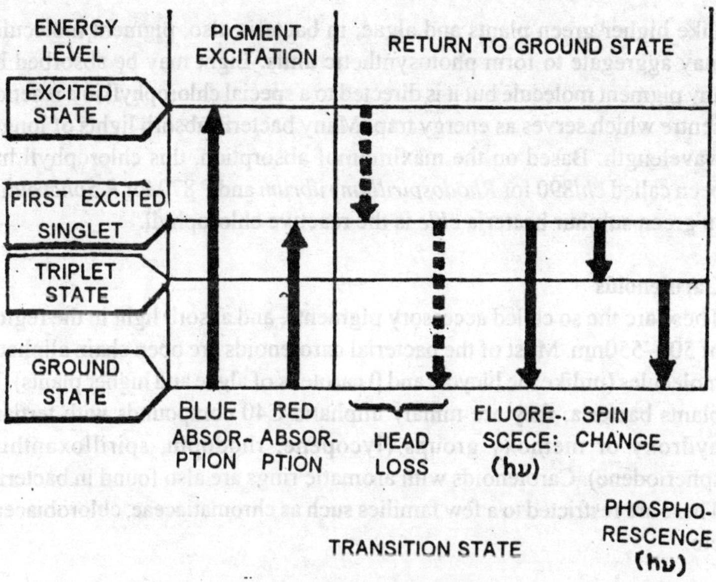

Fig. 5.14 Photosynthesis

Diagram illustrating the absorption of light and changes in the electronic configurations in chlorophyll

Bchl (Bacteriochlorophyll) *a* found in purple sulphur bacteria has a peak between 850–890nm. *Bchl b* which is found in *Rhodopseudomonas viridis, Ectothiorhodospira halochloris* and *thiocapsa pfennigii* has an absorption peak at 1020–1040nm.

Bchl a of purple bacteria occurs in four spectral forms – B 800, B 820, B 850 and B 870–890. The differences in absorption spectra are caused by the kind of binding and the position of the *Bchl* molecule in the pigment protein complexes of the photosynthetic apparatus. The number of peaks and the height varies between species and also cultural conditions.

Chlorobium chlorophyll

These are present in green sulphur bacteria of the family chlorobiaceae. Three chlorophylls– chlorobium chlorophylo 660, chlorobium chlorophyll 650 and chlorobium chlorophyll 715 have been repored in green sulphur bacteria such as *Chlorobium, Chloropseudomonas, Chloroflexus* etc.

According to the recent nomenclature, however, chlorobium chlorophylls of three types mentioned above are called *Bchl c, Bchl* and *Bchl e.*

Like higher green plants and algae, in bacteria also, pigment molecules may aggregate to form photosynthetic units. Light may be absorbed by any pigment molecule but it is directed to a special chlorophyll in a reaction centre which serves as energy trap. Many bacteria absorb lights of longer wavelength. Based on the maximumof absorption, this chlorophyll has been called *chl*890 for *Rhodospirillum rubrum* and P 870 for *R.Spheroides.* In green sulphur bacteria *chle* is the reactive chlorophyll.

Carotenoids

These are the so called accessory pigments, and absorb light in the region of 500–550nm. Most of the bacterial carotenoids are open chain aliphatic molecules (unlike the bicycle and β carotens of algae and higher plants). In plants bacteria, they are minaly aliphatic C40 compounds with tertiary hydroxy or methoxy groups (lycopene, rhodopin, spirilloxanthin, spheriodene). Carotenoids with aromatic rings are also found in bacteria. These are restricted to a few families such as chromatiaceae, chlorobiaceae etc.

Green bacteria possess monocyclic γ carotenoids. Lycopene is the most widely distributed carotene in bacteria. Bicycle carotenes are rare in photosynthetic bacteria. The table below gives various groups of carotenoids found in bacteria. Carotenoids fulfill two functions. On the one hand, they are photosynthetic on the other, they protect the chlorophylls from possible damage due to photoxidative action. It has been seen that carotene free mutants of purple bacteria can grow only in dim light and are killed by higher light intensities.

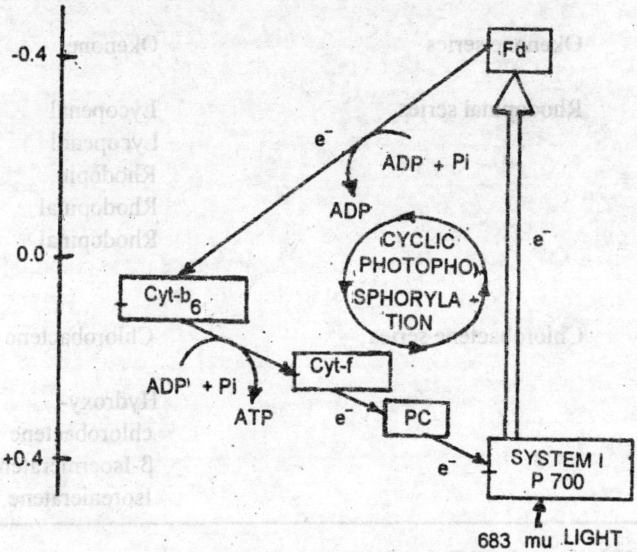

Fig. 5.15 Photosynthesis
Systematic representation of cyclic photophosphorylation

The differences in the absorption spectra of green algae, cyanobacteria, green bacteria and purple bacteria leads to the conclusion that different phototrophic organisms use different spectral rays for their photosynthetic activities. The differences correlate well with the light conditions in their natural habitats. These differences can be used for selective culture of these organisms

Carotenoid groups of phototrophic bacteria

Group	Name	Major components
1.	Normal aspirilloxanthin series	Lycopene
	Rhodopin	
	Spirilloxanthin	
2.	Alternative spirilloxanthin series and ketocarotenoids of spheriodenone type	Spheroidene Hydroxy-spheriodene

	Spheriodenone Hydroxyspheriodenone Spirilloxanthin	
3.	Okenone series	Okenone
4.	Rhodopinal series	Lycopenal Lycopenol Rhodopin Rhodopinal Rhodopinal
5.	Chlorobactene series	Chlorobactene Hydroxy-chlorobactene β-Isoernieratene Isorenieratene

Types of Photosynthetic Bacteria

The prokaryotic phototrophs include a number of Autotrophic pigmented organisms. These are –

1. Purple non sulphur bacteria - Rhodospirillaceae
2. Purple sulphur bacteria - Chromatiaceae
3. Green sulphur bacteria - Chlorobiaceae Chlofoflexaceae
4. Cyanobacteria - formerly blue green algae.

Types of Bacterial Photosynthesis

Phototrophic bacteria normally utilize CO_2 of the environment as a source of cellular carbon. The energy required for the reduction is obtained from inorganic or organic sources. In photolithotrophic organisms, inorganic substances like H_2S or thiosulphate are the source of reducing power. In photoorganotrophic organisms, organic substances like malate or succinate yield reducing power. The general equation for reduction of carbon in photosynthesis is as follows :

$$2H_2A + CO_2 \text{———} C(G_2O) - H_2O + 2A.$$

In algae and other green plants, which grow in the presence of air H_2A is H_2O. In obligatory anaerobic photosynthesizing bacteria. H_2A is H_2S or organic molecules. In chromatiaceae and chlorobiaceae, reductive power comes from H_2S or thiosulphate. In Rhodospirillaceae organic molecules provide reducing power. The following is an outline of different types of reactions involved in bacterial photosynthesis.

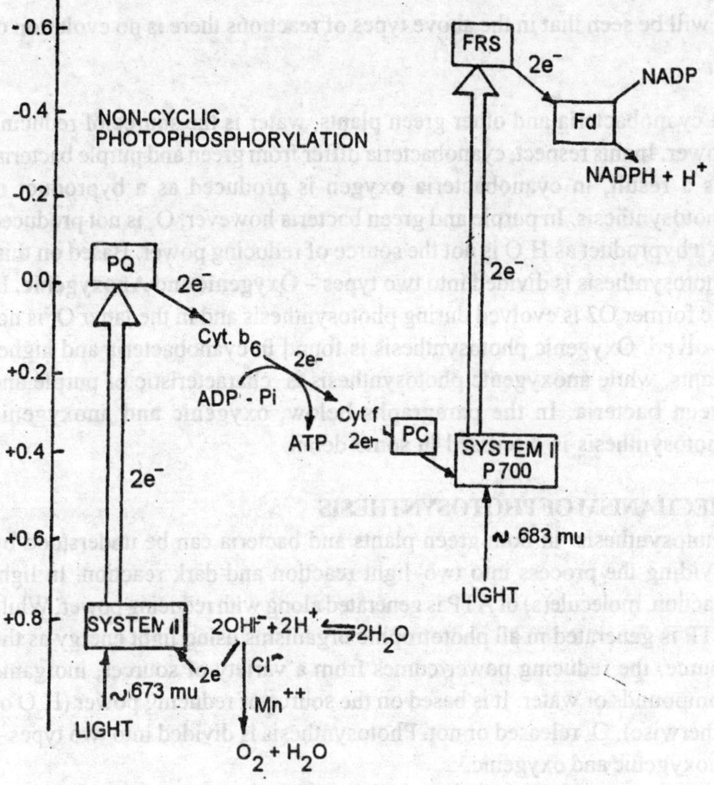

Fig. 5.16 Photosynthesis

Systematic representation of non-cyclic photophosphorylation

1. Purple sulphur bacteria

Here, H_2S is the substrate that provides reducing power. Sulphur is produced instead of oxygen.

$$CO_2 + H_2S \text{———} C(H_2O) + 2S + H_2O$$

Purple sulphur bacteria can also utilize thiosulfate as the source of reducing power

$$2CO_2 + Na2S2O_3 + 5H_2O \text{———} 2C(H_2O) + 2NaHSO_4$$

2. Non sulphur bacteria

Here, the reducing power is provided by organic compounds like ethanol, succinate etc.

$$CO_2 + 3C_2H_5OH \longrightarrow C(H_2O) + CH_3CHO + H_2O$$

It will be seen that in the above types of reactions there is no evolution of O_2.

In cyanobacteria and other green plants, water is the source of reducing power. In this respect, cyanobacteria differ from green and purple bacteria. As a result, in cyanobacteria oxygen is produced as a byproduct in photosynthesis. In purple and green bacteria however, O_2 is not produced as a byproduct as H_2O is not the source of reducing power. Based on this, photosynthesis is divided into two types – **Oxygenic** and **Anoxygenic.** In the former O2 is evolved during photosynthesis and in the latter O_2 is not evolved. Oxygenic photosynthesis is found in cyanobacteria and higher plants, while anoxygenic photosynthesis is characteristic of purple and green bacteria. In the paragraphs below, oxygenic and anoxygenic photosynthesis is discussed in some detail.

MECHANISM OF PHOTOSYNTHESIS

Photosynthesis in both green plants and bacteria can be understood by dividing the process into two–light reaction and dark reaction. In light reaction, molecule(s) of ATP is generated along with reducing power. While ATP is generated in all phototrophic organisms using light energy as the source, the reducing power comes from a variety of sources, inorganic compounds or water. It is based on the source of reducing power (H_2O or otherwise), O_2 released or not. Photosynthesis is divided into two types – anoxygenic and oxygenic.

Light reaction in anoxygenic photosynthesis

Purple sulphur, and green sulphur bacteria conduct anoxygenic photosynthesis. Light reaction here involves the following steps.

1. Source of electron
2. The electron transport system
3. Photophosphorylation

Source of electron

Evolution of oxygen has not been demonstrated in bacteria, and hence these are incapable of using water as a source of reducing power. The source of electron are therefore, hydrogen sulphide and reduced inorganic and organic compounds.

In sulphur bacteria oxidize sulphide to sulphate and the reducing power is utilized to reduce CO_2 as follows :

$$CO_2 + H_2S \longrightarrow 3\,CH_2O + 2S$$
$$3CO_2 + 2S + 5H_2O \longrightarrow 3CH_2O + 2H_2SO_4$$

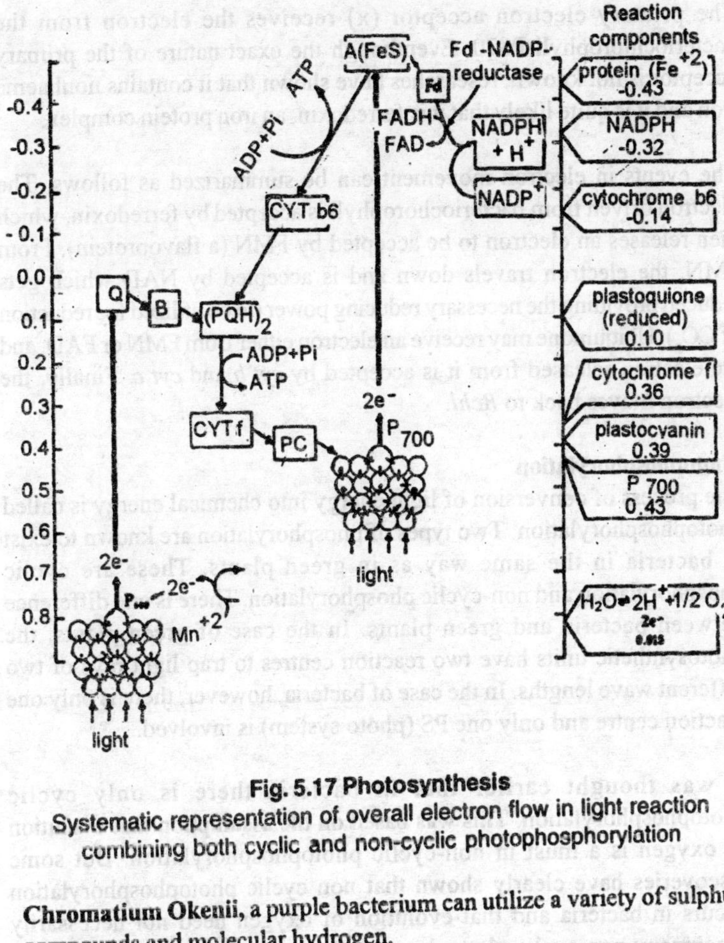

Fig. 5.17 Photosynthesis

Systematic representation of overall electron flow in light reaction combining both cyclic and non-cyclic photophosphorylation

Chromatium Okenii, a purple bacterium can utilize a variety of sulphur compounds and molecular hydrogen.

Electron transport system

In photosynthetic bacteria, the electron transport system consists of a primary electron acceptor (x), a secondary acceptor (y), and *b* and *c* type cytochromes. Cytochrome *a* is absent as molecular oxygen does not participate in the electron transport. Some scientists believe that there are two separate electron pathways–one for photosynthesis and another for respiration. But as is seen in *R.Rubrum,* only one electron transport mechanism is considered enough for both the processes.

The electron transport carriers are asymmetrically located in the membrane to help maintain the hydrogen ion gradient.

The primary electron acceptor (x) receives the electron from the bacteriochlorophyll P870. Eventhough the exact nature of the primary acceptor is not known, researches have shown that it contains nonhaeme iron and it is quite likely that it is ferredoxin, an iron protein complex.

The events in electron movement can be summarized as follows. The electron driven from bacteriochorophyll is accepted by ferredoxin, which then releases an electron to be accepted by FMN (a flavoprotein). From FMN, the electron travels down and is accepted by NAD which gets reduced providing the necessary reducing power (to be utilized for reduction of CO_2). Ubiquinone may receive an electron either from FMN or FAD, and the electron released from it is accepted by *cyt b* and *cyt c*. Finally, the electron moves back to *Bchl*.

Photophosphorylation

The process of conversion of light energy into chemical energy is called photophosphorylation. Two types of phosphorylation are known to exist in bacteria in the same way as in green plants. These are cyclic phosphorylation and non-cyclic phosphorylation. There is one difference between bacteria and green plants. In the case of green plants, the photosynthetic units have two reaction centres to trap light rays of two different wave lengths. In the case of bacteria, however, there is only one reaction centre and only one PS (photo system) is involved.

It was thought earlier that in bacteria there is only cyclic photophosphorylation. This was based on the assumption that evolution of oxygen is a must in non-cyclic photophosphorylation. But some discoveries have clearly shown that non cyclic photophosphorylation occurs in bacteria and that evolution of oxygen need not necessarily accompany non-cyclic photophosphorylation.

Works of Losada, Whatley, and Arnon (1961), have shown that non-cyclic photophosphorylation reaction does not depend on oxygen evolution. The difference between cyclic and non-cyclic photophosphorylation is based on the dependence of the latter on the presence of an electron donor and acceptor which is necessary, produce the reducing power. Occurrence of non-cyclic photophosphorylation in the chromatophores of *Rhodospirillum rubrum* has been shown by suppressing cyclic photophosphorylation with actinomycin *a* and supplying DPIP ascorbate as an electron system, and NAD as electron acceptor.

In bacterial pigment system also, the photosynthetic units consists of two parts namely: an *antenna* consisting of light harvesting molecules and the reaction centre where the actual energy conversion takes place. Light energy harvested by the antenna molecules is channelised to the reaction centre. In the green and purple photosynthetic bacteria about 40 or more bacteriochlorophyll molecules make up the photosynthetic unit. The reaction centre complex is commonly refered to as P870 or P890, although the wavelength band of peak absorption might vary in different species.

In composition, the reaction centre generally consists of *Bchl a* or rarely *Bchl b.* In the purple non sulphur bacterium, *Rhodospirillum rubrum,* the principle light harvesting pigment is *Bchl a* and the reaction centre molecule is P890. In *Rhodopseudomonas sphaeriods,* the light harvesting *Bchl* absorbs at a peak of 850nm and reaction centre *Bchl* is P870.

In green sulphur bacteria, the principle light harvesting pigment is *Bchl c* with a peak 650nm or Bchl *d* with a peak of 660nm. Some of the brown species of green sulphur bacteria have *Bchl e* in the harvesting pigment. The reaction centre chlorophyll is *Bchl a* in both the green and brown species.

The following components are found in the reaction centres of *Rhodopseudomonas sphaeriodes, R.Virids* and *Rhodospirillum rubrum.*

1. Four molecules of *Bchl* a
2. Two molecules of Bacteriopheophytin (*Bph*)
3. 1 molecule of carotenoids
4. One or more quinones
5. One atom of non haeme iron

The reaction centres in most of the photosynthetic bacteria contain three chains polypeptides having a molecular rate of 20,000, 24,000 and 30,000. The *Bchl.* Bacteriopheophytin and Darotenoids are located on the first two polypeptides.

In the primary photochemical reaction, the light quantum absorbed by the light harvesting pigments are transferred to the reaction centre (P870). In the reaction centre, the absorbed photons rais the chlorophyll to excited state and the electron is expelled from the molecule. The P870 now becomes positively charged. The charged P870 transfers the electrons to an intermediate electron acceptor which is reduced. This reduced acceptor then transfers its electron to another acceptor and ultimately the electron comes back to chlorophyll bringing the P870 to its normal state.

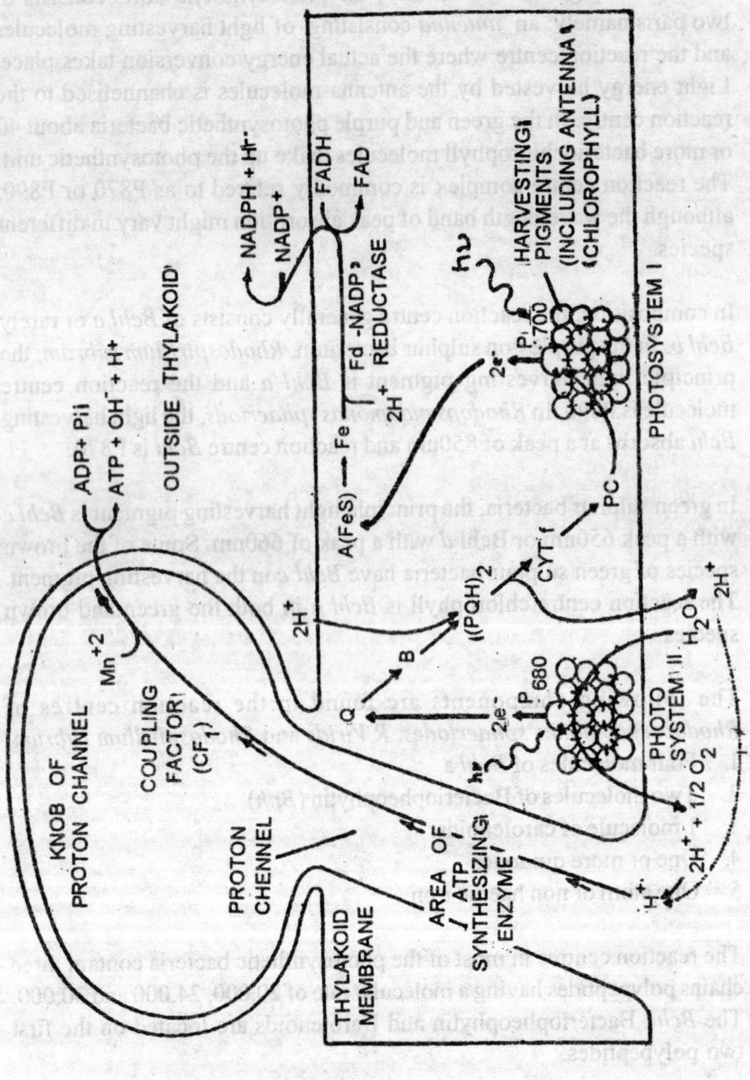

Fig. 5.18 Photosynthesis
Photophosphorylation and ATP synthesis in granum thylakoid according to Chemiosmotic hypothesis

Cyclic photophosphorylation

The process of electron vibration here is the same as what has been explained above. When the electrons are moving from the ferrodoxin to cytochromes, a molecule of ATP is synthesized (ADP + Pi = ATP). This is called cyclic photophosphorylation, because the electron comes back to the chlorophyll in a circular fashion. The net turn of the cycle yields one molecule of ATP.

Non-cyclic photophosphorylation

Here the electron from ferrodoxin instead of moving down towards cytochromes is accepted by NAD, and the reaction is catalysed by ferrodoxin reductase. The reduced NAD now provides the necessary reducing power to fix the carbon dioxide.

The *Bchl* P890 now is in an excited state as it has donated its electron to ferrodoxin. In order to come to the normal state, it must receive an electron from some other source. The electron comes from the oxidation of organic acids, thiosulphates or H2S. These compounds get oxidised as a result of the elimination of an electron. The electron thus released moves via cytochromes back to *Bchl* P890. The chlorophyll molecule now comes back to the normal state and is ready to receive photons from light.

The scheme of photophosphorylation is called non-cyclic because the electron that comes back to *Bchl* is not from its own source but by the reduction of other compounds such as H_2S etc.

Light reaction in oxygenic photosynthesis

The light reaction or photochemical decomposition of water takes place in the grana region of the chloroplast. According to the current understanding, light reaction (Photochemical excitation) takes place in about 10^{-9} seconds. Within this short span of time, there will be the synthesis of molecule(s) of ATP and a molecule of NADPH. These two are utilized in the dark fixation of CO_2. In this section we will study the following aspects in detail with respect to light reaction.

(a) Source of oxygen released during light reaction
(b) Hill's reaction
(c) Arnon's work
(d) Quantum requirement
(e) Emerson effect and two pigment system
(f) Primary photochemical reaction (mechanism of pigment excitation)
(g) Non-cyclic Photophosphorylation.
(h) Cyclic Photophosphorylation

(i) Pseudocyclic photophosphorylation and

(j) Mechanism of ATP formation

SOURCE OF OXYGEN

The overall reaction of photosynthesis upto 1930 was believed to be the exact opposite of respiration and oxygen evolved during photosynthesis came from carbon dioxide.

$6CO_2 + 6H_2O —— C_6H_{12}O_6 + 6O_2$

The discovery by Van Neil (1931), that in certain purple photosynthetic bacteria photosynthesis takes place in the presence of CO_2 but without releasing oxygen. Surely, if the source of O_2 were to be CO_2 even in the absence of water O_2 should have been released. These photosynthetic bacteria used H_2S as hydrogen donor instead of water. Accordint to Van Neil (1931), in purple bacteris H_2S, splits releasing H to reduce CO_2 and in the process elemental sulphur is being formed as follows :

$CO_2 + 2H_2S ——— (CH_2O) + H_2O + 2S$

Van Neil (1931) further argued that, if elemental sulphur comes from H_2S in purple bacteria, in green plants which use water, O_2 should come from water only and not from CO_2. In order to account for the release of O_2 from water the equation should be modified as follows

$6CO_2 + 12H_2O ——— C_6H_{12}O_6 + 6H_2O$

According to Van Neil (1931), water is split up by light (photolysis of water) into H^+ and OH^- ions. The H + ions are used to reduce CO_2, while the OH ions react with one another to form water and oxygen as follows :

$4H_2O —— 4H + 4(OH)^-$

$4(OH)^- —— 2H_2O + O_2$

$CO_2 + 4H^+ —— (CH_2O) + H_2O$

Here (CH_2O) represents not the actual photosynthetic product but the empirical formula of a bigger molecule (glucose).

The work of Ruben, Kamen and Randall (1941), using radio active oxygen (O_{18}) clearly showed that O_2 came only from water and not from CO_2. When the experimental plant was supplied with labelled water (H_2O_{18}), the released oxygen was the O_{18} type, and when the plant was supplied with labelled CO_2 18, the O_2 released was the normal type (O_{16}).

HILL'S REACTION

Hill (1937), and Hill and Scaribrick (1940), showed that the initial reaction of photosynthesis can take place even in isolated chloroplasts in the cell-free environment. When isolated chloroplasts (obtained from ground leaves) are supplied with light, water and a suitable hydrogen acceptor, O_2 will be released and the H_2 acceptor gets reduced. A similar release of O_2 may be observed, when grana from disintegrated chloroplasts are suspended in water with ferric salts and benzoquinones acting as hydrogen acceptors. Besides these, chromates, indophenols etc. also can act as H_2 acceptors. The various H_2 acceptors used by Hill are called Hill oxidants. Hill attributed the release of oxygen to the splitting of water into H and OH by light. Hill's reaction proves the following :

(a) The entire photosynthetic activity including the reduction of CO_2 takes place within the chloroplasts, and

(b) O_{21} released during light reaction comes entirely from water.

ARNON'S WORK (production of assimilatory power)

The products of light reaction provide the necessary inputs for the reduction or assimilation of CO_2. That the assimilatory power (input) is utilized for the dark fixation of CO_2 was demonstrated clearly by the work of Arnon (1951).

According to Arnon (1951), the hydrogen released by water was accepted by coenzyme II (NADP). He found that in isolated chloroplasts, the release of O_2 takes place at a faster rate in the presence of NADP. Arnon also opined that ATP molecules will also be synthesized and he termed this as photophosphorylation.

$$\text{ADP} + \text{Pi} \xrightarrow[\textit{(inorganic phosphate)}]{\textit{energy}} \text{ATP}$$

Arnon also showed that CO_2 fixation takes place in the pigment free portion of the chloroplast (stroma), while the assimilatory power (ATP)/molecules and NADPH), is produced in the grana portion.

QUANTUM REQUIREMENT

The solar radiation that reaches the earth consists of two portions – the visible (400nm – 800nm), and UV, and infrared (less than 400 nm and more than 800 nm). Whether the light consists mainly of waves or particles, as far as photosynthesis is concerned, it is obsorbed in units of energy called photons or quanta.

The quantum yield may be explained on the basis of molecules of CO_2 reduced and amount of O_2 released. Consequently, a minimum quantum is the one required to fix one molecule CO_2 and release one molecule of O_2.

According to Warburg and and Negelein (1922), the minimum quantum requirement is four i.e. 4 quanta are required to fix one mol of CO_2 and release one mol of O_2. This (requirement) concurs with the fact that every CO_2 molecule requires $2H_2O$ molecules for reduction i.e., 4 hydrogen atoms. Emerson and his co-workers (1938), however opine that eight light quanta is the minimum requirement. This has been further confirmed by the work of Govindjee (1968), on *Chlorella* plants. Thus two quanta of light are required for the transfer of every atom of hydrogen during photosynthesis.

EMERSON EFFECT AND TWO PIGMENT SYSTEMS

Emerson and Lewis (1943), worked on different wavelengths of light (monochromatic light) as to their quantum yield and came to the conclusion that 8 quanta of light are required for the reduction of one molecule of CO_2 (or to release one molecule of O_2). Accordingly, the quantum yield is 1/8 or 12%. Since 4 electrons are required in the release of one molecule O_2, Emerson opined that 2 light quanta are required to excite one electron. Proceeding further, Emerson *et al* (1943), determined the quantum yield of different beams of monochromatric light and found that the average quantum yield (12%) suddenly dropped near the far red end of the spectrum.

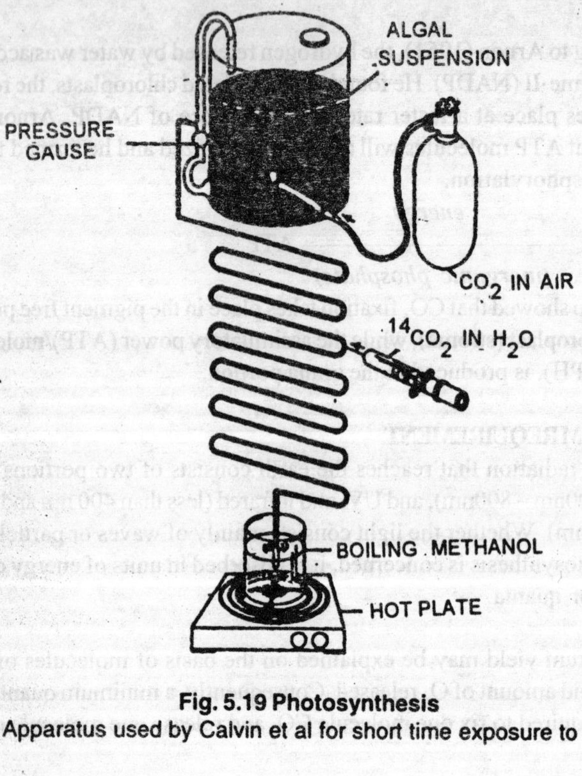

Fig. 5.19 Photosynthesis
Apparatus used by Calvin et al for short time exposure to C^{14}

This decline in quantum yield is called red drop. Red drop begins at wavelengths greater than 680 nm in green plants and greater than 650nm in red algae. The phenomenon of red drop was also confirmed by the works of Haxo and Blinks (1950), on the red alga *Porphyra neregtis.*

The phenomenon of red drop was studied further by Emerson and Chalmers (1958), who found out that the reduction in the rate of photosynthesis due to red drop can be rectified by providing a beam of monochromatic light of shorter wavelength. In other words, by providing shorter wavelengths of light either simultaneously or quickly alternating with a longer wavelength of light, the rate of photosynthesis can be enhanced (effect of red drop is reversed).

This enhancement of photosynthetic rate under the influence of two wavelengths (long and short) of light is called Emerson effect. This may be represented as follows :

$$F = \frac{\text{Rate of evolution of } O_2 \text{ in double beam - Rate of Evolution } O_2 \text{ in red beam (short)}}{\text{Rate of evolution of } O_2 \text{ in far red beam (long)}}$$

Reason for Emerson effect

The discovery of the Emerson effect led to the idea that there must be two different pigment systems in the plants which differed in their absorption peak (of light). While one pigment system was more efficient in longer wavelength, the other was efficient in shorter wavelength.

Obviously, under a specific wavelength (either short or long), the rate of photosynthesis will not be maximum as one or the other pigment, system will be ineffective. Butler (1966), discovered two forms of chlorophyll viz *chl a* 673 and *chl a* 683 (with absorption peaks at 673nm and 683 nm respectively). Clayton (1966), discovered another form of *chl a* with absorption peak at 700nm which he named P700.

It has now been confirmed by various researches that two pigment system PS I and PS II are involved in photosynthesis, each pigment system having different absorption peaks and consisting of *chl a* and other accessory pigments. The two pigment systems may be removed from the thylakoids, by treating the chloroplast with mild detergents. These can be separated on polyacrylamide gel when they appear as discreet green bands. The details of the two pigments system are as follows :

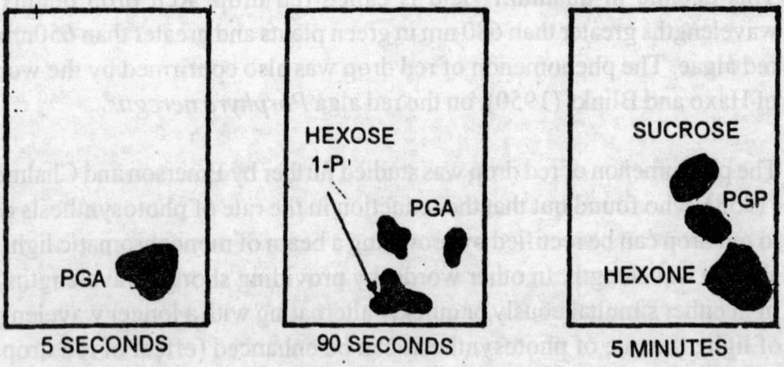

Fig. 5.20 Photosynthesis
Auto radiograph (Diagrammatic representation of chromatograms showing the fixation of C^{14} after different periods of exposure to light

Pigment System (PS I)

According to Malkin (1982), Anderson (1982), and Barber (1983), PS I consists of Chl *a* 670, Chl *a* 680 and Chl *a* 710 (P 700), Chl *b* and some β cartoenes. P700 is the reaction centre in PS I into which light energy is transferred by all other pigments. Besides, PS I also consists of iron containing proteins called Fe-S proteins and Cytochromes. PSI is weakly fluorescent.

Pigment System II (PSSII) :

It has several forms of Chl *a* with P680 (Chl a 680) being the reaction centre. A little amount of Chl *b* and β carotenes are also present. There is also a colourless Chl *a* called Pheophyrin (Klima and Krasnovskii, 1981), which acts as the primary electron acceptor together with these are present quiones (Q) and manganese proteins.

Light Harvesting complexes

Electrophoretic separation of pigments from Chloroplasts, reveals the presence of two other green bands besides PS I and PS II (Salisbury and Ross, 1986). Each band contains Chl *a* + Chl *b* but very little β carotene. All these pigments are protein bound. According to Salisbury and Rosss (1986), these green bands represent light harvesting complexes of pigments and proteins; one band functions in association with PS I and the other with PS II. The function of the light harvestng complexes is to absorb light energy and to transfer it to the appropriate pigment system.

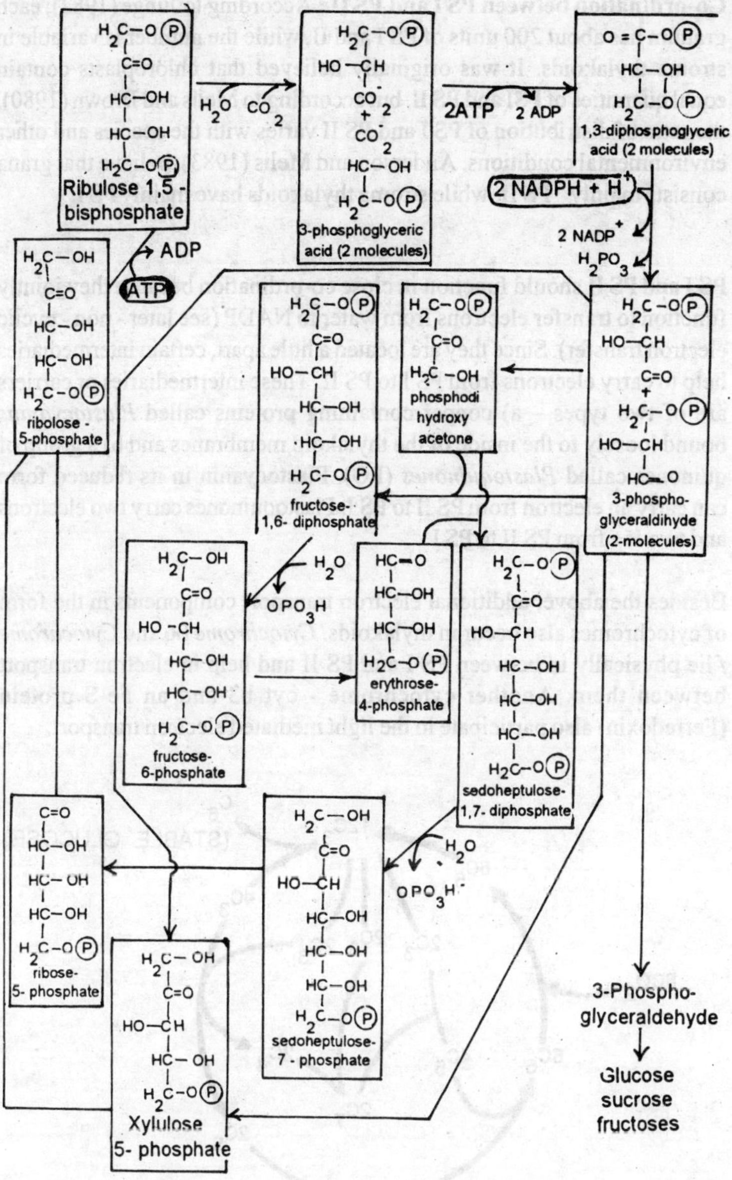

Fig. 5.21 Photosynthesis
Systematic representation of C3 pathway (Calvin Cycle

Co-ordination between PS I and PS II : According to Junge (1982), each granum has about 200 units of PS I and II, while the number is variable in stroma thylakoids. It was originally believed that chloroplasts contain equal quantities of PSI and PS II, but according to Melis and Brown (1980), the ratio of distribution of PS I and PS II varies with the species and other environmental conditions. Anderson and Melis (1983), believe that grana consists mainly PS II, while stroma thylakoids have mainly PS I.

PS I and PS II should function in close co-ordination because they jointly function to transfer electrons from water to NADP (see later - non - cyclic electron transfer). Since they are located a little apart, certain intermediaries help to carry electrons from PS I to PS II. These intermediaries or carriers are of two types – a) copper containing proteins called *Plastocyanins* bound loosely to the inside of the thylakoid membranes and b) a group of quinones called *Plastoquinones* (PQ). Plastocyanin in its reduced form can carry an electron from PS II to PS I. Plastoquinones carry two electrons and two H+ from PS II to PS I.

Besides the above, additional electron transport components in the form of cytochromes also occur in thylakoids. *Cytochrome* b6 and *Cytochrome* *f* lie physically in between PS I and PS II and help in electron transport between them. Another cytochrome - cyt b3 and an Fe-S protein (Ferredoxin) also participate in the light mediated electron transport.

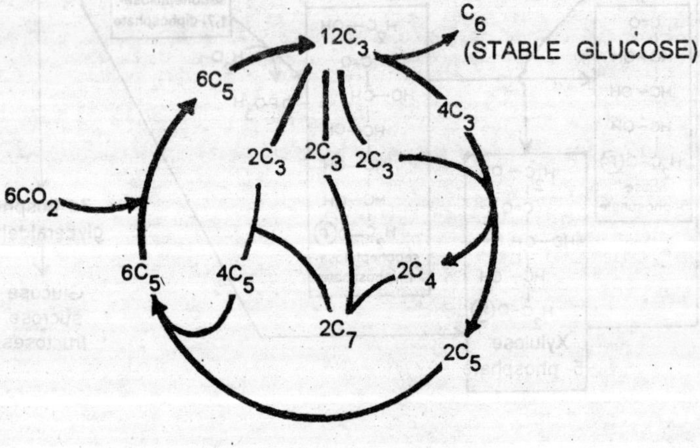

Fig. 5.22 Photosynthesis
Balance Sheet of carbon Calvin's Cycle

Also present in thylakoids and necessary for light mediated ATP synthesis is, a group of complex proteins called *coupling factor* (CF) complexes of ATP ases (see later under mechanism of ATP formation for more details). The CF complex consists of a spherical head attached to the thylakoid membranes towards the stroma and a stalk that extends across the lipid bilayer.

MECHANISM OF PIGMENT EXCITATION

Normally, the chlorophyll molecule wil be in its stable state (no excitation), in the sense the two electrons of its atoms will be spinning in opposite directions and the molecule has no magnetic movement.

When the chlorophyll molecule absorbs light energy in the form of photon one of the electrons is raised to a higher orbit (higher energy). However, the spin of the electrons continues to be in opposite directions, and hence to magnetic movement. The unexcited state of the chlorophyll molecule (before the absorption of light) is called the *ground state* or the *single state.* Here, the energy level is zero, hence it is also called the E_0 state. After the excitation, the stage is known as the first excited state or the S_1 state, with the energy being E_1. The life span of the S_1 state is very short and very soon (10^{-9} sec), the excited electron returns to the ground state by losing energy in the form of heat, light, or by the transfer of energy. The chlorophyll molecule reaches the S_1 state by absorbing the light at the red level (660 nm). If the light is absorbed at the blue level (400 nm), the chlorophyll molecule isknown to reach the S_2 level (with energy level being E_2), because the 'blue photon' has more energy. However, even here the magnetic movement is nil as the spin of the electrons is still in the opposite directions. The electron of S_2 state reaches the ground state via the S_1 state by losing energy either as heat or as heat of light (fluorescence or phosphorescence). Since the electron reaches the ground state very soon (10^{-8} or 10^{-9} see), the energy cannot be put into use (it canot be converted into chemical form).

Under certain circumstances, an electron at the S_1 state may lose some energy in reversing the spin and drops into a low orbit. At this state, there will be two electrons in two different orbits, but spinning in the same direction. The molecule is now said to possess magnetic movement with electrons oriented in three different ways relative to the axis. This is called the *triplet state* or the *metastable state* (T_1 state). The conversion of S_1 state electron to T_1 state is called *Inter system crossing.*

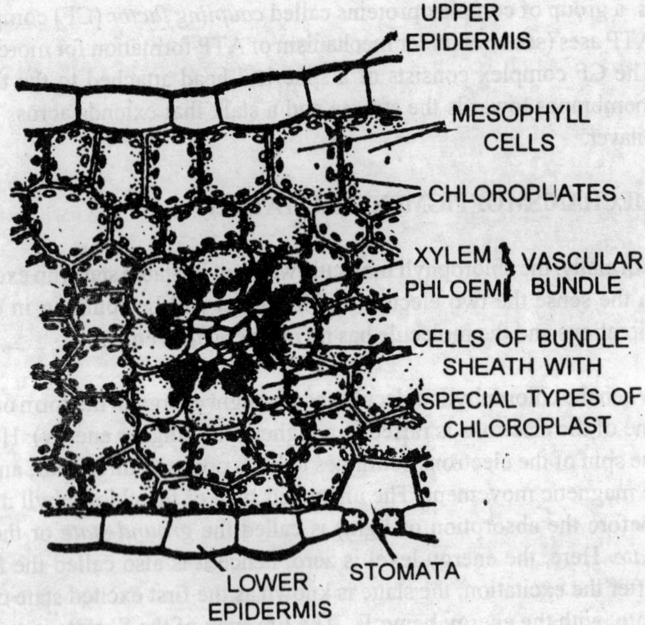

Fig. 5.23 Photosynthesis

TS of a monocat leaf (grass) showing Kranz anatomy

The lifespan of T_1 state electron is longer (10^{-2} or 10^{-3} sec). The electrons do not dissipate energy quickly, for they have to reverse their spin before reaching the ground state. Since the electron in the T_1 state remains energised for a long time, the energy can be made use of for chemical reaction (conversion into chemical energy). The electron at the T_1 state is dislodged from the chlorophyll molecule and is shuttled between a series of electron acceptors, the energy is utilized to produce ATP and (reduced) NADPH molecules. The series of electron acceptors constitute the electron transport system (ETS).

Photophosphorylation

Arnon and Frankel (1954), showed that isolated chloroplasts, under illumination can produce ATP molecules. The reaction is believed to take place as follows.

$$ADP + Pi \xrightarrow[\text{Chlorophyll}]{\text{Light}} ATP$$

(ADP = Adenosine Diphosphate)
ATP = Adenosine Diphosphate)
Pi = Inorganic phosphate

Aron *et al* (1954), named this phenomenon as Photophosphorylation (addition of phosphate group to ADP under the influence of light energy). These ATP molecules are utilized to reduce PGA to PGAL during the fixation of CO_2. They are also required to phosophorylate Ribulose 5 phosphate to Ribulose 1.5 diphosphate (For details, see dark reaction). The following types of photophosphorylation have been discovered in plants.

1. Cyclic Photophosphorylation
2. Non - Cyclic Photophosphorylation, and
3. Pseudocyclic Photophosphorylation

Cyclic Photophosphorylation

Here only PS I participates. The light energy absorbed by the light have harvesting complex which is transfered by inductive resonance to P700. P700 can accept light energy only of a wavelength longer than 680 nm. The excited electron of PQ700 is first accepted by an unknown acceptor (x), then to ferredoxin. This causes ferredoxin to release one of its own electrons (energy rich), which shuttles through the chain consisting of Cyt b6, Cyt f. state and now ready to receive one more unit of radiant energy.

The electron released from the chlorophyll (P700) is returned in a continuous chain i.e., it travels in cyclic fashion, hence the name, cyclic electron transport or cyclic photophosphorylation.

Synthesis of ATP during cyclic photophosphorylation

The electron ejected from P700 has an energy of 0.42 v which gets reduced to 0.417 when it reaches ferredoxin. Here the electrons passes through 3 down hill migration (Cyt b6, Cyt f and Pi) steps before it is returned to P700. The potential gap between Ferrodoxin to P700 (ground state) is about one volt which is sufficient to produce at least two molecules of ATP. ATP molecules are formed at two stages – a) between ferredoxin and Cyt b6 and b) between Cyt b6 and Cyt f

In cyclic photophosphorylation, water does not participate; hence there is neither evolution of O_2 nor is there any formation of reduced NADPH molecule (for there is no hydrogen donor). Further, PS II is not involved in any stage.

According to Park and Sane (1970), there are two types of PS I, PS I*s* and *PS1g*. *PS 1s* is present in the stroma thylakoids and *PS 1g* is present in the granum. *PS 1s* participates in cyclic photophosphorylation, while *PS 1g* participate in non-cyclic photophosphorylation together with PS II.

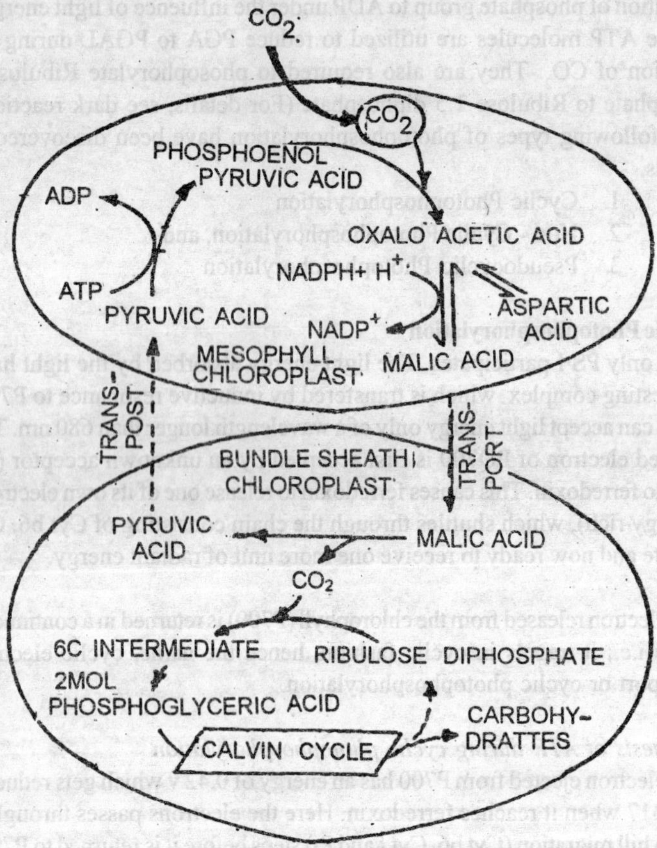

Fig. 5.24 Photosynthesis
Systematic representation of Hatch and Slack pathway

Ramreiz et al (1968), opine that cyclic photophosphorylation is of limited occurrence in plants, and a high rate of cyclic reaction may even retard the CO_2 fixation as NADPH will not be available. They argue that cyclic reaction is not involved in the main path of carbohydrate synthesis as it can only provide part of ATP requirement for CO_2 fixation.

Non-Cyclic Photophosphorylation

This is the major pathway of light reaction involving both PS I and PS II. Non-cyclic photophosphorylation is also called the 'Z' scheme electron transport because of the zig-zag fashion of electron travel. It takes place as follows :

(a) The radiant energy is trapped by PS I (P700), which ejects one if its electrons and is trapped by X, FRS, and finally goes to ferredoxin. The reduced ferredoxin now transfers its electrons to NADPH. NADPH gets reduced. PS I, however remains in the excited state as it has not got back its electron.

(b) Meanwhile, PS II absorbs light energy (680nm) and ejects one of its own electrons which is trapped by an unknown quinone acceptor (Q). From Q–the electron moves downhill to cyt, b6, plastoquinine, cyt f, Plastocyanin (PC), and finally to PS I. PS I which is in an excited state, now returns to the ground state as it has got an electron from PS II.

How does PS II come back to the ground state now that it has sent its electron to fill up the 'hole' in PS I?. Because, unless the PS II returns to the normal state, the system cannot contunue functioning. The answer lies in the dissociation of water molecules. Water splits into H^+ and OH^+ ions. What is the mechanism of this splitting of water? Some physiologists call it photolysis of water, others, the breakdown of water. In any case, the excited PS II returns to the normal state by getting an electron from the OH^- ions of water. According to Salisbury and Ross (1986), the excited PS II returns to the ground state by attracting an electron from an adjacent Mn^- protein, which in turn gets an electron from the OH^- of water. The H^+ ions released by water are accepted by NADPH to become NADPH + H^+. Thus, a reduced NADPH is formed.

The electron transport system or photophosphorylation is called non-cyclic because the excited PS I returns to the normal state by getting an electron from PS II, while the excited PS II returns to the ground state by receiving an electron from water. Hence, the movement of electron is not in a cyclic fashion.

ATP Synthesis during non-cyclic photophosphorylation

The electron (+ 0.8 v) released by PS II is accepted by a quinone from where the downhill migration begins to reach PS I via Cyt b6, Cyt f and PC. One molecule of ATP is synthesized when the electron shuttles between PQ and Cyt f.

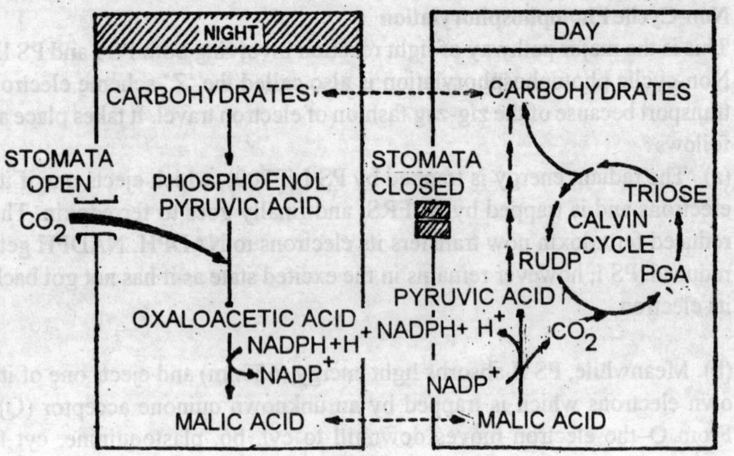

Fig.5.25 Photosynthesis
Systematic representation of CO_2 fixation in CAM Plants

Release of oxygen and formation of NADPH + H+

Four molecules of water split up for every turn of non-cyclic photophosphorylation. This is necessary (see fig) to provide four electrons to the excited PS II to bring it back to the ground state. Similarly, four H + ions are required to reduce 2 molecules of NADPH.

These reactions take place as follows :

$4H_2O$ —— $4H^+ + OH^-$

$4OH^-$ ——— $2H_2O + O_2$

(4 electrons from OH reach the excited PS II)

2NADPH + — $4H^+$ ——— 2nadph + $2H^+$

(four electrons are required to reduce 2 molecules of NADPH and these come) from the excited PS II via ferredoxin)

Thus the end products of non-cyclic photophosphorylation are :

(a) One molecule of ATP
(b) Two molecules of reduced NADPH
(c) 2 molecules of water and
(d) One molecule of O_2.

While a and b are used for the dark fixation of CO_2, c and d are byproducts.

Comparison between cyclic and non-cyclic photophosphorylation

	Cyclic	Non-cyclic
1.	Only PS I participates	PS I and PS IIparticipatre
2.	PS I gets back the electron in a cycle fashion	PS I gets back the electron from PS II from water
3.	Only ATP molecules are formed	Both ATP and NADPH molecules
4.	Water does not participate, hence no evolution of O_2.	O_2 evolved as water participates
5.	Found predominantly in bacteria	Found predominantly in green plants.

Pseudocyclic Photophosphorylation

Arnon and his co-workers demonstrated another kind of Photosphosphorylation, when illuminated chloroplasts produced ATP molecules in the absence of CO_2 and NADP, but in the presence of FMN or Vit K and oxygen. This involves the reduction of FMN by water with the evolution of oxygen. The reduced FMN is reoxidizable by oxygen. Hence, Arnon (1954), called it oxygen dependent FMN catalysed photophosphorylation or pseudocyclic photophosphoyrlation.

MECHANISM OF ATP FORMATION

The source of energy for the synthesis of ATP molecules from ADP comes obviously from the electron transport chain. There are many evidences which point out to this relationship brought about by some common agents. For instance, if some uncoupling agents are introduced into the system, electron transport continues, while the ATP synthesis is inhibited. If the uncoupling agent is removed, ATP synthesis is resumed. Inhibition of electron transport also inhibiting ATP synthesis is another evidence to show that both are coupled.

While there does not seem to be any doubt about the coupling of electron flow and phosphorylation, the mechanism is as yet not fully understood (according to some physiologists). The following are the three hypotheses proposed for the mechanism of ATP synthesis.

(a) Conformational coupling
(b) Chemical coupling
(c) Chemiosomotic coupling

(The mechanism of ATP synthesis is the same for both photophosphorylation during photosynthesis as well as for oxidative phosphorylation during respiration.)

Conformational coupling

During electron transport, both in mitochondria as well as in chloroplasts, some conformational (structural) changes occur which facilitate the release of energy resulting in the production of ATPase mediated ATP synthesis, ATPase is a reversible action enzyme which decomposes ATP into ADP and Pi under less energy conditions, while facilitates synthesis of ATP during high energy conditions. According to Devlin and Witham (1983), electron micrographs of membranes indicate structural variations during peak activity. But there does not seem to be any definite relationship between activated membrane and ATP synthesis.

Chemical coupling

This hypothesis believes in the presence of a coupling factor (most probably a protein), mediating the energy transfer between electron transport and ATP synthesis. Initially, the coupling factor (CF) forms a high energy CF complex with one of the electron carriers. Later, an inorganic phosphate replaces the electron carrier in the CF complex. This results in the formation of high energy rich phosphate to ADP resulting in the formation of ATP.

Chemiosmotic coupling

Among the hypothesis proposed for the mechanism of ATP synthesis, this is by far the most acceptable, Originally proposed by Michell during 1960s for oxidative phosphorylation in mitochondria, it has recently been applied to photophosphorylation in chloroplasts also (Jagendorf, 1975).

As it relates to oxidative phosphorylation, the flow of electrons from NADH to atmospheric oxygen through the respiratory electron transport, influences the extrusion of protons from the mitochondrial matrix. As a result, a proton concentration gradient is produced across the mitochondrial membrane. In effect, the energy of electron flow in the respiratory chain is partially conserved as the *proton concentration gradient* or *proton motive force* across the inner mitochondrial membrane. The protons return to the mitochondrial matrix down the electrochemical gradient through ATPase resulting in the formation of ATP from ADP and Pi.

Ubiquinone, one of the components of the respiratory chain known to transfer protons across the mitochondrial membrane through alternate oxidation and reduction. The flow of protons across the membrane is due to a force called *Proticity* (Noggle and Fritz, 1983), which is comparable to electricity. According to Noggle and Fritz (1983), the word chemiosmotic was proposed to emphasize the fact that chemical energy (oxidation energy) is converted to osmotic energy (difference in the proton concentration gradient between the inner and outer faces of the membrane).

Oxidative phosphorylation may be halted by treating the membrane with uncouplers. These are weak acids which dissolve in the protoplasmic membrane and disturb the proton concentration gradient by allowing the free flow of protons. Thus the proton uncouplers (protonophores) disturb the ATP synthesis.

The Chemiosmotic theory also explains ATP formation in chloroplats. Jagendorf (1975), demonstrated that a pH gradient established across the thylakoid membranes stimulates the ATP synthesis. The electrons are transported across the membrane to the interior through the electron transport chain. Light energy provides the necessary drive the flow of electrons from water to NADP and for transport of protons across the membrane. Plastoquinone, an electron acceptor transfers electrons to Cyt f and picks up H + ions on the outside of the thylakoid membrane and releases protons to the interior of the channel. This results in the building up of a pH gradient (the interior of the thylakoid becomes acidic due to accumulation of protons) across the membrane. The accumulated protons flow from the inside to the stroma side of the membrane through pathways of coupling factor (which include stalks and knobs), which are the sites of photophosphorylation. The downhill migration of protons provides the necessary energy for the addition of Pi to ADP to form ATP, with ATPase acting as a mediator.

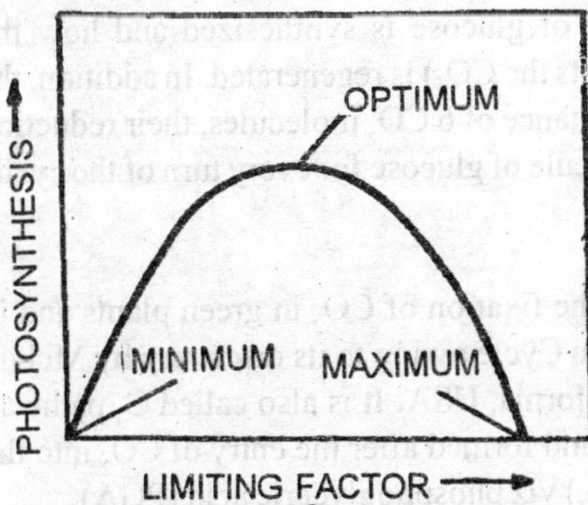

Fig. 5.26 Photosynthesis
Sach's concept of 3 cardinal points

DARK REACTION OF PHOTOSYNTHESIS (carbon assimilation)

The source of carbon for reduction is atmospheric CO_2 for green plants. For photosynthetic bacteria, carbon may come from Co_2 or from organic compounds. The process of CO_2 assimilation in photosynthetic bacteria is of the following types :

1. Reductive pentose pathway or Calvin cycle
2. Reductive carbosylic acid cycle or pyruvate synthetase pathway
3. Organic carbon assimialation pathway

CALVIN CYCLE (reductive pentose pathway)
Calvin cycle or the C_3 cycle, which is the predominant pathway of CO_2 fixation in higher plants is seen in photosynthetic bacteria also. Researches carried out on *Rhodospirillum ruburm* and *Thiobacillus* spp using radio isotopes of carbon (Fog, 1968), have confirmed this. Sugar phosphates are, however, less as these bacteria are unable to utilize or store them.

The phenomenon of thermochemical reduction of CO_2 was first established by Blackmann in 1905. He opined that light is not necessary for the reduction of CO_2. In dark reaction the products of light reaction (ATP and NADPH + H) are used to reduce CO_2 into glucose. But the precise manner in which CO_2 is absorbed, and reduced resulting in the formation of a molecule of glucose was understood fairly recently, thanks to the availability of radioactive isotopes. Utilizing the isotopes of carbon dioxide, one can trace the different stages of entry of CO_2 into the plant system untill it gets converted into glucose. Many physiologists, biochemists etc. have contributed a great lot towards the understanding of the path of carbon in Photosynthesis. The names of Bassham (1957 - 1964), Melvin Calvin (1961), Benson (1951) etc are worth mentioning in connection with the untravelling of the secrets of CO_2 fixation. Melvin Calvin was awarded the Nobel Prize in 1961 for tracing the CO_2 reduction cycle, which is also known as the Calvin cycle.
In CO_2 reduction cycle, what we have to precisely understand is how the CO_2 enters the plant system, which is the compound that accepts the CO_2 molecule and how a molecule of glucose is synthesized and how the compound (that originally accepts the CO_2) is regenerated. In addition, the system should explain the acceptance of 6 CO_2 molecules, their reduction and formation of only one molecule of glucose for every turn of the cycle.

CALVIN CYCLE
This is the major pathway for the fixation of CO_2 in green plants and in some bacteria. It is called Calvin Cycle owing to its discovery by Melvin Calvin of the university of California, USA. It is also called C_3 pathway because, the first stable compound formed after the entry of CO_2 into the plant system is a triose sugar (C_3) viz phosphoglyceric acid (PGA).

Some experimental details of Calvin's work
In order to understand the significance of the discovery of CO_2 reduction

cycle, we should understand the experimental methods followed by Calvin.

To trace the various steps of conversion of CO_2 after its entry into the plant system until the formation of glucose, Calvin allowed the experimental plants to photosynthesize for various periods of time and analysed the various compounds formed after different periods of photosynthesis.

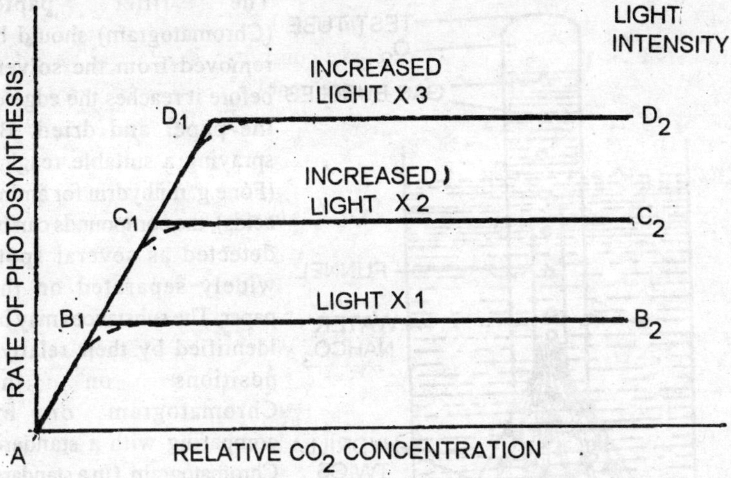

Fig. 5.27 Photosynthesis
Graph representing Law of Limiting Factors

Clavin chose for his experimentation two green algae – *Chlorella pyrenoidosa* and *Senedesmus olbiqus.* The plants have the same pigmentation as higher green plants and can be easily cultured and harvested. To make sure that the compounds are already present in the plant, the experimental plants were starved of CO_2, and later supplied with a radioactive isotope of CO_2 (C_{14}). Only those compounds in which C_{14} was incorporated were taken into consideration because they were produced after the experiment began. By varying the time of photosynthesis (1/16 sec to 5 sec), and then killing the plant to stop photosynthesis and analysing the intermediate compounds formed, Calvin and his co-workers were able to map the whole of the CO_2 reduction cycle. Two, very simple but efficient tools is namely *Chromatography* and *Radioauthgraphy,* to trace the path of carbon in photosynthesis.

Chromatography, in particular, paper chromatography is an ideal technique for the separation of compounds present in a small mixture. The extract (mixture) of the plant to be analysed is placed as a spot on paper and a solvent is allowed to run. This solvent due to capillary action carries with it the different substances in the extract. The rate of migration of the differnt substances depends upon their solubility and specific gravity.

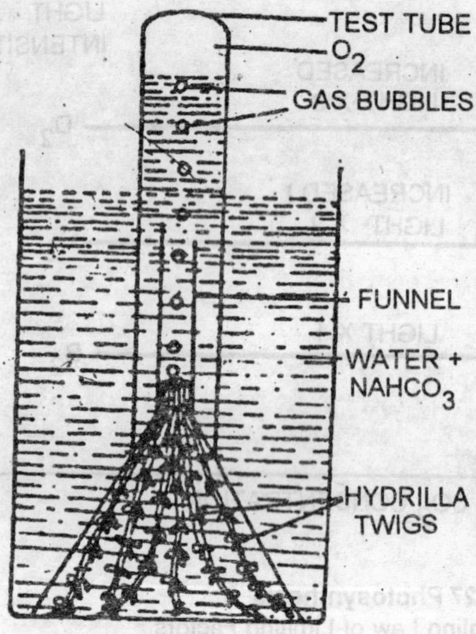

Fig. 5.28 Photosynthesis
Experiment to demonstrate the release of O_2 during photosynthesis

The filter papter (Chromatogram) should be removed from the solvent before it reaches the edge of the paper and dried. By spraying a suitable reagent (For e.g. ninhydrin for amino acids), the compounds can be detected as several spots widely separated on the paper. The substances may be identified by their relative positions on the Chromatogram or by comparing with a standard Chromatogram. (In a standard Chromatogram, compounds whose identity are known are spotted, and their position ismarked after the solvent run).

Radioautography is a technique to find out, if a particular source consists of a particular type of isotope. Different radioactive isotopes emit different types of radiation. Chiefly, there are 3 types. These are : (a) Alpha rays (b) Beta rays (c) Gamma rays.

When we use a particular we already know what is the type of emission. For instance, C_{14} emits β rays. When the rays are directed to a photographic film they from characteristic markings, and by these one can identify the type of radiation and thereby the isotope.

In the present expirement, Melvin Calvin and associates allowed the algae to photosynthesise for varying periods (using C_{14}), after which they (plants) were killed in boiling methanol. The plant extract was then spotted on the Chromatogram to identify the compounds. In order to make sure that these contain C_{14}, the chromatogram was held against a photographic film of equal size. Only those compounds which had incorporated C_{14} leave markings on the film and these were taken into consideration (as produced during experimentation). In order to vary the time of photosynthesis, flashes of light from 1/16 sec to 5 sec were given to identify the earliest compound formed. In this way, Calvin and his associates found out all the intermediaries in the CO_2 reduction cycle.

The first stable compound formed after the initial entry of CO_2 is a triose (PGA). The triose cannot be the acceptor molecule for it has to initially, have a 2C compound to accept one CO_2 molecule to become triose. Benson, however, soon found out that a 5 carbon compound is the first acceptor molecule (Ribulose 1.5 diphosphate). After accepting a molecule of CO_2, an unstable 6C compound is formed, which immediately breaks up into 2 molecules of PGA (triose), the first stable compound.

The essentials of the Calvin cycle are as follows. The cycle should be self exclusive in that it should remain unaltered, but provide for the continuous entry of 6 molecules of CO_2, formation of one molecule of glucose and regeneration of six molecules of Ribulose 1.5 diphosphate (RUDP) 1, the initial acceptor molecule. The important events of CO_2 cycle can be detailed as under.

(a) Fixation of CO_2
(b) Reduction of PGA
(c) Formation of sugars, and
(d) Regeneration of RUDP

Fixation of CO_2

CO_2 is first accepted by molecules of a pentose (5C) sugar Ribulose 1.5 diphosphate, and an unstable 6 carbon compound is formed. This compound (tentatively called carboxy dimsutase system) immediately breaks up into 2 molecules of a triose (3C) called phosphoglyceric acid (PGA). The reaction is catalyzed by the enzyme carboxy dismutase of Rubp carboxylase (Rubisco). For every molecule of CO_2, 2 mols of PGA are formed; hence 12 PGA moles for $6CO_2$ mols.

Reduction of PGA

The molecules of PGA are reduced to phosphoglyceraldehyde (PGAI) by NADPH + H. The reaction requires energy and hence. ATP participates. The enzyme triose phosphate dehydrogenase catalyses this reaction.

Rudp + CO_2 H_2O ——— PGA (2 molecules)

ATP

PGA + NADPH + H ——— PGAI

The above reaction takes place at two stages. In the first stage, PGA + ATP Phosphoglycerokinase 1,3 - diphosphoglyceric acid + ADP In the ssecond stage.

1.3 Diphosphoglyceric acid + NADPH + H —— 3 PGAI + NADP

Since 6 molecules of CO_2 are taken into the system (by 4 mols of Rudp), 12 PGAL mols are formed after reduction of the 12 molecules of PGA. Only two are utilized for the formation of sugars, while the rest are channeled back for the regeneration of Rudp molecules. Several intermediate compounds are formed in the process.

Formation of sugars

The PGA molecule (one out of two) undergoes isomerization to form a mol of dihydroxyacetone phosphate. The enzyme triose phosphate isomerase catalyses this reaction. Ultimately, sugars are formed as follows :

Triosephosphate

1. 3 PGAL ——— Dihydroxyacetone phosphate

Isomerase

2. 3 PGAL + Dihydroxyacetonephosphate

Aldolase

——— Fructose 1.6-diphosphate

Fructose 1,6-diphosphate now gets converted into fructose 5 monophosphate due to the loss of one phosphate group under the influence of the enzyme phosphotase.

Phosphatase

3. Fructose 1.6-dip ——— Fructose 6. Monophosphate (FMP).

+H_2O

Fructose 6 monophosphate isomerises into glucose monophosphate under the influence of the enzyme isomerase.

4. EMP Isomerase Glucose 6 monophosphate (GM)

Now FMP or GMP may undergo dephosphorylation to form either fructose or glucose. These monosaccharides unite to form sucrose and various other carbohydrates.

Regenration of Rudp

In the above explanation of formation of sugars, it has been pointed out that Rudp accepts a molecule of CO_2 to ultimately form a molecule of sugar. But, the overall formula for photosynthesis accounts for the entry of 6 molecules of CO_2, hence, there must be 6 molecules of Rudp to accept the

CO_2 molecules and for every turn of the cycle, this should result in the formation of 6 molecules but there is only one molecule of glucose. Then what happens to the remaining 6 molecules of hexose (5moles of 6C). Obviously they undergo transformation to regenerate 6 molecules of Rudp. This is necessary format Rudp to be ready to accept and be self-sufficient; and there must always be 6 molecules of Rudp ready, to accept 6 molecules of CO_2 and form 6 molecules of glucose, the cycle will cease to operate for the next turn of the cycle there will be no Rudp to accept CO_2 molecules. Thus for the entry of every 6 CO_2 molecules, the output is one mol of hexose and the rest go for the regeneration of Rudp.

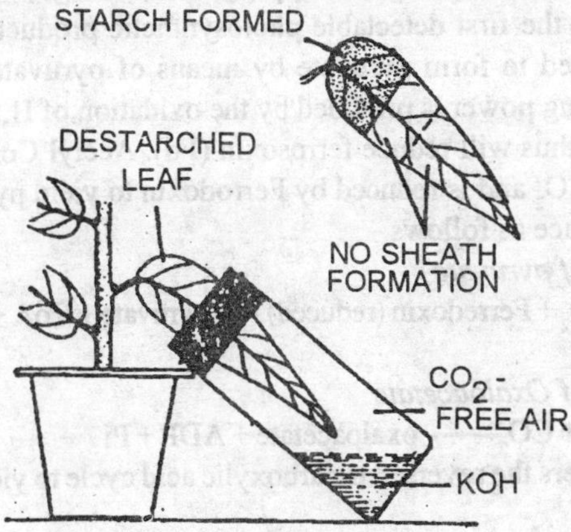

Fig. 5.29 Photosynthesis
Experiment to demonstrate the necessity of light for photosynthesis

The balance sheet of the molecules of carbon from the entry of CO_2 to the generation of Rudp is given in Fig.

In the following equations, regeneration of one mol of Rudp is explained (and this may be multiplied to obtain 6 molecules of Rudp).

1. Fructose 6 Phosphate + PGAL $\xrightarrow{\text{Transketolase}}$ Erythrose4, Phosphate+ Xylulose 5. Phosphate
2. Erythrose 4, Phosphate + Dihyrosxyacetone phosphate $\xrightarrow{\text{Aldolase}}$ Sedophetulose 1.7. Diphosphate

3. Sedoheptulose 7. Phosphate + PGAL $\xrightarrow{\text{Phosphatase}}$ Xylulose 5 H_3PO_4.

4. Sedoheptulose 7. Phosphate + PGAL $\xrightarrow{\text{Transketolase}}$ Xylulose 5 Phosphate + Ribose 5 Phosphate

Xylulose 5 Phosphate and Ribose 5 phosphate get converted into Ribulose 5 phosphate under the influence of the enzyme phosphophentose isomerase

5. Ribulose 5 Phosphate $\xrightarrow[\text{-Kinase}]{\text{Phosphopento}}$ Rudp

Reductive carboxylic acid cycle

This has been demonstrated in green bacteria such as *Chlorobium thiosulfatophilum* and *Chromatium* spp (Evans et albugo, 1966). In this type of fixation, the first detectable photosynthetic products are amino acids. CO_2 is used to form pyruvate by means of pyruvate synthetase reaction. Reducing power is provided by the oxidation of H_2S. Reducing power obtained thus will reduce ferrosoxin (Fd). Acetyl CoA (of Krebs cycle), accepts CO_2 and is reduced by Ferrodoxin to yield pyruvate. The reactions take place as follows –

1. *Formation of pyruvate*

Acetyl CoA + CO_2 + Ferredoxin (reduced) —— pyruvate + CoA + Ferredoxin (oxidized)

2. *Formation of Oxaloacetate*

Pyruvate + ATP + CO_2 —— oxaloacetate + ADP + Pi

Oxaloacetate enters the reversed tricarboxylic acid cycle to yield succinyl CoA.

3. *Formation of α Ketoglutarate*

Succinyl CoA + CO_2 + Ferredoxin (reduced) —— α Ketoglutarate + CoA + Ferredoxin (Oxidized)

4. *Formation of Ctirate*

α Ketoglutarate is converted into citrate through oxalosuccinate. Citrate the, splits into oxaloacetic acid and acetate.

α Ketoglutarate CO_2 —— oxalosuccinate —— citrate —— oxaloacetic acid + Acetate

The above reactions are generally irreversible, but because of string reducing potential of the Ferrodoxin, the reaction can be reversed. As these bacteria can generate reduced ferrodoxin by means of photosynthesis, it is called reductive carboxylic acid cycle (Evans, Buchanan, Arnon, 1966).

The net result for every turn of the cycle is that four molecules of CO_2 are incorporated (reductive fixatin), and on molecule of oxaloacetate is synthesized. Besides ferredoxin, the cycle requires all other cofactors of

Krebs cycle in reduced form. Three molecules of ATP ar required as follows for this fixation -

1. Activation of acetate 2. Carboxylation of pyruvate and
3. Activation of succinate.

Organic carbon assimilation pathway

Certain types of bacteria use organic compounds to provide CO_2 for fixation (not atmospheric CO_2) Stainer (1961), opined that a part of the organic carbon is directly assimilated. Accordingly to her, photoassimilated organic substrates (from organic polymers) such as acetate, is directly converted into poly β hydroxybutyrate as follows :

According to Gaffron (1963), however, during photoassimilation CO_2 reduction occurs only when excess reductants are available. Under anaerobic conditions, photometabolism of organic acids is strictly light dependent, and may be dependent on light generated ATP.

COMPARISON BETWEEN BACTERIA AND GREEN PLANTS (in terms of photosynthesis)

In the photosynthetic activity, there are a number of similarities and a few differences between green plants and photosynthetic bacteria. It is believed that photosynthesis in green plants is evolved from that found in simple green and purple bacteria. There is difference between green plants and bacteria in the localization of the photosynthetic pigments; while well organized chloroplasts are there in the former, in the latter, pigments are localized in membranous vesicles. (There is infact an opinion that the chloroplasts with their own DNA can be regarded as some primitive phototrophic organisms which have been trapped by cells of plants in the course of evolution).

The photosynthetic pigments viz., the chlorophylls of bacteria are different from those of higher plants in terms of their absorption spectra. It should be pointed out, however, that the basic chemical framework of all chlorophylls is the same. In carotenoid composition, however, green plants and bacteria differ–open aliphatic chain Carotenoids are seen in the latter, while the former have bicyclic Carotenoids.

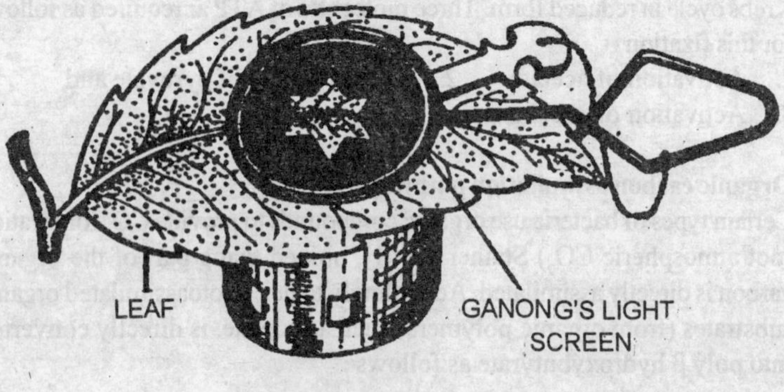

Fig. 5.30 Photosynthesis
Experiment to demonstrate the necessity of light for photosynthesis

In terms of the photochemical processes, bacteria are *anoxygenic,* while green plants (Cyanobacteria also) are *oxygenic.* This is due to the fact that in bacteria, the reducing power comes from H_2S and other inorganic and organic compounds, while in green plants water gets reduced and oxygen is released in the process. It is an enigma as to why bacteria have not chosen H_2O as a source of reducing power.

The source of cellular carbon is always atmospheric CO_2 in green plants, while it may be CO_2 or other organic compounds in bacteria.

There are a number of similarities also between bacteria and green plants in pigment composition. *Chl a* is present in both, though bacteriochlorophyll has two more hydrogen than green plant chlorophyll. Both chlorophylls are formed in membrane bound particles in the cell. In both, chlorophyll systems reaction centres exist with antenna pigments channelising light energy for harvesting.

In both green plants and bacteria, the overall photosynthesis is an oxidation reduction reaction. A reductant H_2A gets oxidized and provides the reducing power to convert carbon to cellular carbon. Oxidized 'A' is released in the process. Cyclic and non-cyclic photophosphorylation is found in both. Electron transport chain leading to the formation of ATP and reduced NAD compound is found in both green plants and bacteria. Carbon assimilation is similar as well as different in both the groups with Calvin cycle being common.

There is some difference in the photoreductant compound. It is $NADPH_2$

in green plants and $NADPH_2$ in photosynthetic bacteria. But the chromatophores of bacteria have an enzyme transhydrogenase and can photoreduce NADP in the presence of NAD. The following are the summarized similarities and differences between bacterial photosynthesis and green plant photosynthesis.

Similarities

1. Presence of chlorophyll *a*
2. Arrangement of photosynthetic pigments in membranous particles in the cell.
3. Presence of a definite reaction centre for harvesting light energy.
4. Accessory pigments channelise light energy towards reaction centre.
5. Occurrence of photophosphorylation, cyclic as well as non-cyclic.
6. Electron carriers of the chain are almost similar.
7. Light energy produces the reducing power.
8. Overall photosynthesis is an oxidation–reduction process.

Differences between photosynthesis found in bacteria and green plants

Bacteria	Green Plants
1. Bacteria have no distinct chloroplast	Chloroplast is well developed
2. Bacteria absorb light of longer wavelength (800-900 nm or infra red)	These absorb light of relatively shorter wavelength (450-700 nm).
3. P_{890} is the reaction centre	P_{680} and P_{700} are two reaction centres
4. Chlorophyll *a* is absent and bacteriochlorophyll takes its functions	Chlorophyll *a* is present which converts radiant into chemical energy.
5. The carotenoids are open chain alephatic type	Carotenoids are bicyclic
6. Oxygen is not evolved during the product (anoxygenic photosynthesis)	Oxygen is evolved as by product(oxygenic photosynthesis).
7. Water does not serve as a source of reducing power (electron donor)	Water serves as source of reducing power.

8.	Bacteria can use CO_2 as well as organic compounds as a source of carbon	Only CO_2 is the source of carbon
9.	The photoreductant is $NADH_2$	The photoreductant is $NADPH_2$.
10.	The process occurs in the presence of light but in the absence of O_2	Both are present during the process
11.	Emerson effect is not found	Emerson effect is found dominant
12.	Cyclic photophosphorylation is dominant	Non-cyclic photophosphorylation isdominant
13.	Plastocyanin is absent	Plastocyanin is present
14.	Lycopene is found higher	Lycopene is not found in the chloroplast of plants.

6

RESPIRATION

The most basic requirement of all living organisms is *energy.* Without energy, no life activity is possible. The solar energy that is trapped by green plants are be stored in cells in the form of carbohydrates and other nutrients. These are oxidised and the energy so released utilized to carry on the vital processes of living beings.

The process of energy release by the (oxidation of) substrate material is called respiration. The term respiration, which has been in usage since a long time is derived from the latin word *respirare* (= breathe). Breathing as a phenomenon of life has been noticed in animals long before its relation to respiration and its universal occurrence in all forms of life.

Breathing and respiration

Breathing and respiration are often used as synonymous. But it is not correct. Breathing as a separate process is noticed only in few organisms (animals), while respiration occurs in all organisms. Further, breathing is a physical process, while respiration is a chemical process taking place in all the cells. Breathing may be regarded as the last step found only in aerobic respiration, where atmospheric oxygen participates. There can be respiration without breathing, but the reverse is not possible.

Respiration and combustion

Respiration is often compared to the process of combution, because in both the instances the substrate material gets burnt (oxidized) up. But the process of respiration differs from mere combustion in many respects. In combusion, the burning up or oxidation of the substrate is sudden, while in respiration it is stepwise. In combustion, a large amount of energy is wasted as heat, while in respiration it is channelised and stored up for further usage. Hence, respiration may be termed as an orderly combustion found in living organisms.

Cellular respiration

In a multicellular organism where there are thousands of cells, how does energy reach each cell? Each cell obtains its own energy by the oxidation of substrate material present within. Hence, respiration takes place in all the cells wherever they may be present in animals and plants. This is called cellular respiration.

Cellular respiration is a complex process which includes various aspects. These are the following.

(a) Absorption or oxygen (not in all)

(b) Degradation of substrates such as carbohydrates to CO_2 and H_2O (oxidation).

(c) Energy is released in the process of oxidation; some amount will also be lost in the form of heat.

(d) Formation of many intermediate products which play different roles in metabolism.

(e) There will be some loss of weight due to the process of oxidation. From the point of view of plant metabolism, cellular respiration is of paramount significance as it is the basic biological process that provides energy for all the vital activities. In short, the following aspects may be mentioned as indicating the significance of respiration.

  (i) Energy released in respiration is used for various metabolic processes. In fact, most of the physiological activities are driven by respiratory energy.

  (ii) Many intermediate compounds are formed which play a key role in the intermediate metabolism.

  (iii) Insoluble food is converted into soluble form.

  (iv) Potential energy is converted into kinetic energy.

  (v) CO_2, which is released during respiration is an important step in maintaining the carbon balance of nature.

The overall formula for respiration is as follows :

$$C_6H_{12}O_6 + H_2O + 6O_2 \quad 6CO_2 + 12\,H_2O + 686\text{ K cal energy.}$$

Respiratory substrate

The energy rich material (food substrate), that is used for the purpose of oxidation is called the *respiratory substrate*. In other words a respiratory substrate is defined as "any organic plant constituent oxidized partially or completely with the concomitant release of energy together with CO_2 and H_2O".

Plants use a variety of organic constitutents as their respiratory substrates. These may be a variety of carbohydrates (hexose sugars, disaccharides, polysaccharides), organic acids, fats etc. Under certain circumstances like starvation, even proteins may be used as respiratory substrates.

Among the carbohydrates, sucrose and starch are the principal respiratory substrates. When simple monosaccharides like glucose and fructose are available, they are directly utilized as respiratory substrates, while sucrose and starch are first hydrolysed to simple sugars before being channeled into respirtory metabolism.

In certain seeds like castor bean, fats are reserve food materials. They are first broken down into fatty acids and glycerol. As already mentioned, proteins will be channeled into respiratory metabolism under special circumstances such as starvation. Blackmann, who has studied respiration in detail, observed that in leaves kept in darkness, the respiratory rate which will be normal initially, declines, but does not stop altogether. Blackmann termed the normal respiration (when carbohydrates are utilized), as *floating respiration* and the slowed down rate of respiration as *protoplasmic respiration,* when the essential proteins are oxidised to keep the cells alive. In the long run, however this is harmful to the cells. When proteins are the respiratory substrates, they are first broken down into amino acids and these by deamination get converted into keto acids and enter the respiratory chain.

Respiratory Quotient (RQ)

RQ may be defined as *"the ratio between the volume of carbon dioxide given out and oxygen taken in simultaneously, by a given weight of the tissue in a given period of time, at standard temperature and pressure".*

RQ may be calculated as follows

$$RQ = \frac{\text{Volume of } CO_2 \text{ evolved}}{\text{Volume of } O_2 \text{ consumed}}$$

The value of RQ depends upon the nature of the respiratory substrate, amount of oxygen present (in the substrate), and the extent to which the substrate gets broken. Normally, the value of RQ should be unity (1), but many deviations are noticed due to the alterations in the levels of oxidation and reduction. The value of RQ also depends upon whether all the oxygen consumed is used up only for respiration or some of it is utilized for other purposes also. It is for these reasons that RQ may be unity, less than unity or even more than unity. RQ value also gives an idea of the respiratory substrate that is being used. The following are some RQ values when different respiratory substrates are involved.

RQ for Carbohydrates

Carbohydrates are the principal respiratory substances in a large majority of the organisms. These include a variety of monosaccharides (glucose, fructose etc), disaccharides (sucrose) and polysaccharides (starch, inulin etc). When simple sugars are available they are directly, channelised into the respiratory stream, but when poly and disaccharides are involved, they are first hydrolysed into monosaccharides.

When hexose sugars are the respiratory substrate, volume of CO_2 evolved equals volume of O_2 used, and hence the value will be unity. This will be illustrated by the following equation.

$$C_6H_{12}O_6 \rightarrow 6O_2 + H_2O$$

$$RQ = \frac{CO_2}{O_2} = \frac{6CO_2}{6O_2}$$

It has to be noted that not in all cases when sugar is the respiratory substrate, the RQ will be unity. There will be many deviations due to the following reasons.

(i) Incomplete oxidation of sugars (as in anaerobic respiration).
(ii) Involvement of many reductive events such as sulphate and nitrate reduction.
(iii) Oxidation and decarboxylation processes unrelated to respiration.
(iv) Retention of CO_2 within the cell fluids instead of releasing it to the exterior.
(v) Non photosynthetic carboxylation reactions resulting in accumulation of these products as seen in succulent plants.
(vi) Metabolic utilization of CO_2

RQ for Fats

Fats are not generally found in vegetative parts, but they are found as storage products in many of the angiosperm seeds like sunflower, castor, niger seed, groundnut etc. At the time of the germination of seeds, a major amount of fat is converted into carbohydrates, while the rest is used in respiration.

Fats are poor in oxygen content, and the ratio of oxygen to carbon is always less in fats when compared with carbohydrates. Fats are rich in hydrogen and as a result they require more O_2 for complete oxidation. In addition to this, fats are not directly oxidised. They are first broken into *fatty acids* and *glycerol,* which then act as respiratory substrates. Oxygen is required for the breakdown of fats into fatty acids and glycerol. As a result, for the overall respiratory process (when fats are involved), more oxygen is required. Hence, the value of RQ will be less than unity.

(a) $C_{18}H_{36}O_2 + 26O_2 \rightarrow 18\,CO_2 + 18H_2O$
(Stearic acid)

$$RQ = \frac{26\,O_2}{18CO_2} = 0.69$$

(b) $C_{57}H_{16}4O_6 + 80O_2 \rightarrow 57CO_2 + 52H_2O$
(Triolein fat)

$$RQ = \frac{57}{80} = 0.7$$

When simple fatty acids like acetic acids are involved, RQ will be unity as seen below.

(c) $CH_3COOHY + 2O_2 \rightarrow 2CO_2 + 2H_2O$
(Acetic acid)

$$RQ = \frac{2O_2}{2\,CO_2} = 1.\text{ (unity)}$$

(d) $2C_{51}H_{98}O_6 + 145\,O_2 \rightarrow 102\,CO_2 + 98H_2O$
(Tripalmitin)

$$RQ - \frac{102}{145} = 0.7$$

Interestingly, fats liberate approximately three times more energy when compared with Carbohydrates. While a gram of carbohydrates yeilds about 3.8 K cal, the same amount of fat produces approximately 9.1 K cal of energy. This may be the reason as to why the seeds prefer to store reserve food material in the form of oil, more amount of energy can be obtained by less amount of food (it requires less space).

RQ for Proteins

Proteins and amino acids serve as respiratory substrates only under certain situations like starvation etc. Like fats, proteins also have less oxygen and the volume of oxygen to carbon is low. When proteins are hydrolysed, they require more oxygen for complete oxidation as a result of which RQ value will be less than unity. The RQ value for proteins and their derivatives varies between 0.5–0.8; very often it will be 0.79.

The value of RQ may be 0.99 when ammonia is released or it may be 0.79 when amides are formed.

RQ for Organic acids

The value of RQ is more than unity when organic acids are used as respiratory substrates. This is due to the fact that in organic acids, the proportion of oxygen in relation to carbon is always more, or less of O_2 is required for oxidation. The following are some of the examples.

(a) $C_4H6O_5 + 3O_2 \rightarrow 4CO2 + 3H2O$
(malic acid)

$$RQ = \frac{4CO_2}{3O_2} = 1.3$$

(b) $2(COOH)_2 + O2 \rightarrow 4CO_2 + 2H_2O$
(oxalic acid)

$$RQ = \frac{4}{1} = 4.0$$

(c) $2C_4H_6O_6 + 5O_2 \rightarrow 8CO_2 + H_2O$
(tartaric acid)

$$RQ = \frac{8}{5} = 1.6$$

(d) $2C6H8O/ + 902 \rightarrow 12CO2 + 8H2O$
(Citric acid)

$$RQ = \frac{12}{9} = 1.33$$

RQ for succulent plants

In succulent xerophytes such as *Opuntia,* and many other plants (*Bryophyllum, Sedum* etc), carbohydrates are incommpletely oxidized resulting in the formation of intermediate compounds, there being no evolution of CO_2. As only O_2 is taken in and there is no release of CO_2, RQ is zero.

In succulent plants, stomata are open during night, O_2 is taken in with the result carbohydrates are incompletely oxidized, while during day time, the organic acids are completely oxidized releasing CO_2. But this CO_2 is not released as it is used for photosynthesis. Hence, with no evolution of CO_2 RQ will be zero.

$$2C_6H_{12}O_6 + 3O_2 \rightarrow 3C_4H_6O_5 + 3H_2O$$

$$RQ = \frac{CO_2}{O_2} = \frac{O}{3} = 0.0$$

RQ when O_2 is used for other activities

O_2 is also utilized for various other processes like converon of fats to carbohydrates, synthesis of anthocyanins etc. In such cases, the amount of CO_2 evolved will not correspond to the amount of O_2 consumed. Here RQ falls below unity.

RQ for anaerobic respiration

Certain tissues respire anaerobically (no participation of atmospheric, O_2), in which case only CO_2 is evolved without the absorption of O_2. In such circumstances, RQ will be more than unity.

$$C6H12O6 \rightarrow 2C2H5OH2 + 2CO2$$

$$RQ = \frac{CO2}{O_2} = \frac{2}{O} = 2$$

The following is a list of RQ values for some of the substrates occuring in various plant parts.

No.	Plant part	Substrate	SQ value
1.	Leaves	Carbohydrates	1.0
2.	*Opunitia* shoots	Organic acids	0.03
3.	Seeds	Starch	1.0
4.	Linsed	Fat	0.64
5.	Buckwheat seed	Protein	0.5
6.	Pea Seeds	Carbohydrates	1.5.2.4

Measurement of RQ

RQ can be measured with the help of *Ganong's respiroscope.* The apparatus essentially consists of a graduated tube with a bulb at the upper end and a levelling tube. The two tubes are connected at their base with a long rubber tube which assumes a 'U' shape. The bulb in the graduated tube has a glass stopper fitted into its neck. Both the neck and the stopper have holes which can be brought into a straight line by twisting the stopper. The apparatus is fitted into an iron or wooden stand, and filled with mercury (instead of Hg, saline solution also can be used). Respirable material such as plant parts or germinating seeds are kept in the bulb and the stopper is rotated to allow the air enter into the bulb. Mercury level in both the tubes is brought at the same level. Now the stopper is rotated to cut off the communication.

The initial level of Hg is noted. If the respiratory substrate is carbohydrate, CO_2 released will be equal to O_2 observed, so that there is neither rise nor decrease in the level of Hg. If the substrate is fat or protein, CO_2 released will be less than O_2 consumed and therefore a vaccum will be created in the chamber. This results in the increase of the level of Hg.

If the CO_2 released is more than O_2 used up, the initial level of Hg will fall after the experiment. RQ in this case will be more than one. RQ may be calculated with the help of the following formulae.

$$RQ = \frac{v^2}{v^1 + v^2}$$

v^1 = Initial rise of Hg; v^2 = further rise in Hg level after introduction of caustic potash. This is understood to find out the excess release of CO_2 if any.

$$RQ = \frac{v^2}{v^2 --- v^2}$$

v^1 = Fall in the initial level of Hg. v^2 = Rise in level after introduction of caustic potash.

TYPES OF RESPIRATION

The involvement of atmospheric oxygen during respiration constitutes the basis for its categorization into aerobic and anaerobic types.

Aerobic respiration

The respiratory substrate is completely broken down into CO_2 and H_2O with the participation of atmospheric O_2.

$$C_6H_{12}O_6 + 6O_2 \rightarrow 6CO_2 + 6H_2O + 686K\ Cal.$$

Anaerobic respiration

Here there is no participation of atmospheric oxygen. Food substances are incompletely oxidized. Hence, all the carbon in the substrate will not be released in the form of CO_2. The end products are not only CO_2, but intermediate products like ethyl alcohol, acids etc.

$$C_6H_{12}O \rightarrow 2C_2H_5OH + 2CO_2 + 56K\ Cal.$$

Besides the above, there are some more differences between aerobic and anaerobic respiration. They are listed below.

	Aerobic	Anaerobic
1.	O_2 necessary	O_2 not necessary
2.	Common to most of the organisms	Restricted to certain organisms, and certain tissues.
3.	Occurs always	Occurs temporarily in certain tissues, besides microorganisms
4.	Oxidation complete	Oxidation incomplete
5.	Large amount of energy is liberated	Less energy is liberated
6.	End Products are CO_2 and water	End products are ethyl alcohol, organic acids etc and CO_2
7.	The entire carbon content of the substrate comes out in the form of CO_2 (inorganicc	Due to incomplete oxidation of the substrate, carbon will still be left in the organic compounds (ethyl alcohol)
8.	Process non toxic to plants	Toxic to plants
9.	Respiratory enzymes present in mitochondria cytoplasm and cytoplasm.	Enzymes present only in
10.	Output in 38 molecules of ATP	Output is only 2 molecules of ATP

MECHANISM OF RESPIRATION

As has already been pointed out, during respiration the substrate material is broken down with release of energy, CO_2 and H_2O.

In the mechanism, we will be studying the various stages of the breakdown of the substrate material, enzymes involved etc., and account for the release of CO_2 and energy released in the form of molecules of ATP. In the foregoing paragraphs, various steps involved in the breakdown of carbohydrates are given.

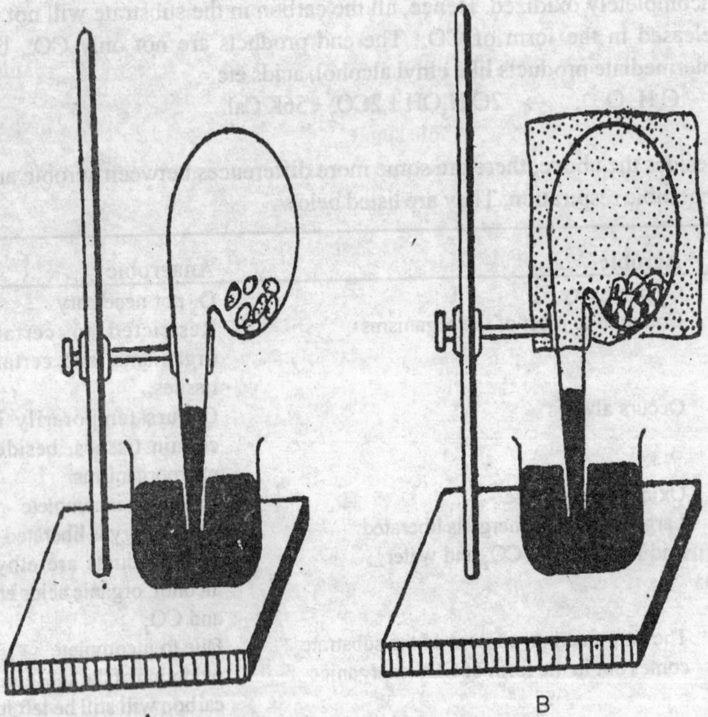

Fig. 6.1 Respiration

Demonstration of value of RQ for carbohydrates. **A.** Initial leel, **B.** Ater 24 hours

Broadly categorizing, the following steps are identifiable in the oxidative breakdown of carbohydrates.

1. Hydrolysis 2. Glycolysis 3. Pyruvic acid oxidation 4. Krebs cycle or citric acid cycle and 5. Electron transport system or oxidative phosphorylation.

The mechanism of anaerobic of anaerobic respiration consists of pyruvic acid.

1. Glycolysis and 2. Anaerobic oxidation of pyruvic acid.

From the above amount it can be understood that the step glycolysis is common for both aerobic and anaerobic respiration.

HYDROLYSIS

When starch is the reserve food material, it should first be hydrolysed into simple sugars before it can enter into the respiratory cycle. The process of breakdown of starch into simple sugars with the addition of water is called hydrolysis.

GLYCOLYSIS

During glycolysis, one molecule of glucose will be broken into two molecules of pyruvic acid. Originally, the name glycolysis was meant for the breakdown of glycogen (*lysis* or breakdown of glycogen) found in muscle cells. Enzymes responsible for glycolysis are present in the cytoplasm. As will be seen later, no participation of atmospheric O_2 is required for the formation of pyruvic acid. Glycolysis is the common pathway for both aerobic and anaerobic respiration.

Glycolytic breakdown of glucose into pyruvic acid is also known as the **Embden - Meyerhof - Parnas pathway (EMP pathway), after three scientists** who discovered the various steps of glycolysis. EMP pathway is also called cytoplasmic respiration as the whole process of glycolysis takes place only in cytosol (cytoplasm minus organeloles). EMP pathway was first worked out in muscle cells and yeast, and later it was observed that it is common for all the organisms. The overall reaction of glycolysis may be summed up as follows.

$$C_6H_{12}O_6 \rightarrow 2CH_3COCOOH + 4H$$

The details of the above reaction are given below.

Phosphorylation

Any sugar that has to enter into the glycolysis pathway must first undergo phosphorylation (i.e., a phosphate group has to be added). In this process, ATP participates and donates a phosphate group (there will be utilization of a molecule of ATP).

(i) Glucose + ATP Glucose 6 Phosphate + ADP

The reaction is catalyzed by the enzyme *hexokinase* with Mg ++ as the cofactor. This reaction was first worked out by Meyerhof in 1927. Glucose 6 phosphate may be produced by another means also, if starch is the original food material. It is as follows.

Starch + H3PO4 → Glucose 1 phosphate

The enzyme phosphorylase catalyses this reaction. In the next step

Glucose 1 phosphate → Glucose 6 phosphate

In the above reaction Glucose 1 phosphate changes into Glucose 6 phosphagte i.e., the phosphate group attached to the first carbon atom is shifted to the sixth carbon atom. The reaction is catalyzed by the enzyme *Phosphoglucomutase.*

Isomerization

In the step, glucose 6 phosphate isomerises into Fructose 6 phosphate under the influence of the enzyme *Phosphohexoisomerase.*

Glucose 6 phosphate → Fructose 6 phosphate

Fructose 6 phosphate may be produced from fructose directly when the latter is available.

$$\text{Fructose + ATP} \xrightarrow[\text{Mg}^{++}]{\textit{Hexokinase}} \text{Fructose 6 phosphate}$$

Formation of Fructose 1.6. diphosphate

This is a key reaction in the initial breakdown of glucose. Fructose 6 phosphate utilizes a molecule of ATP to get converted to Fructose 1.6 diphosphate under the influence of the enzyme *Phosphohexokinase.*

Fructose 6 phosphate + ATP → Fructose 1.6 Diphosphate + ADP

The reactions of phosphorylation leading to the formation of Fructose 1.6 diphosphate may be summarized as follows.

(i) $\text{Glucose + ATP} \xrightarrow[\text{Mg}^{++}]{\textit{Hexokinase}} \text{Glucose 6. Phosphate + ADP}$

(ii) $\text{Glucose 6 Phosphate} \xrightarrow{\textit{Phosphohexoisomerase}} \text{Fructose 6 Phosphate}$

(iii) $\text{Fructose 6 Phosphate + ATP} \xrightarrow[\text{Mg}^{++}]{\textit{Phosphohexokinase}} \text{Fructose 1.6 Diphosphate + ADP}$

Cleavage of Fructose 1.6 dip

Further reactions in glycolysis begin with the breaking up of Fructose 1.6 diphosphate into two molecules of triose under the influence of the enzyme *Aldolase.* The presence of this enzyme in plant tissues was first discovered by Tewfik and Stumf (1949).

1. Fructose 1.6 diphosphate ______ Glyceraldehyde 3 phosphate + Dihydroxyacetone Phosphate.

97% of Fructose 1.6 diphosphate gets converted into glyceraldehyde 3 phosphate, while about 3% will change into dihydroxyacetone phosphate. Dihydroxyacetone phosphate isomerises into glyceraldehyde 3 phosphate under the influence of the enzyme *phosphotriose isomerase*. The reaction may be represented as follows.

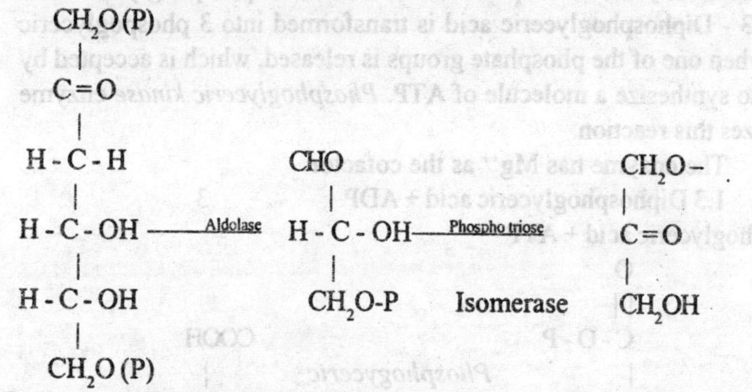

Fructose 1.6 Diphosphate   Glyceraldehyde 3 - Phosphate   Dihydroxyacetone Phosphate

To sum up, Fructose 1.6 diphosphate (6C) breaks up into 2 molecules of glyceraldehyde 3 phosphate (3C). From now on the reactions are explained for a molecule of triose (glyceraldehyde 3 phosphate). While accounting for hexose, each reaction has to be doubled. In the next step, glyceraldehyde 3 phosphate is oxidized (hydrogen removed) with the help of the enzyme *phosphoglyceraldenhyde dehydrogenase*. Inorganic phosphate and NAD participate in this reaction. 2 electrons and two protons (H+) are released from glyceraldehyde 3. Phosphate and the energy is utilized to link the inorganic phosphate to the oxidized glyceraldehyde 3 phosphate to form 1,3 - diphosphoglyceric acid.

Glyceradehyde 3 phosphate + H_3PO_4 + NAD ———— 1.3 Diphosphoglyceric acid + NADH + H +

As may be noticed, while the electrons provide the energy necessary for the linking of inorganic phosphate, the H + goes for the reduction of NAD to NADH +H^+

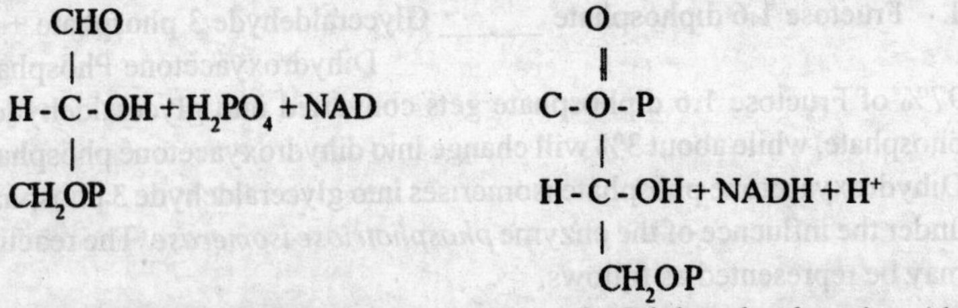

lyceraldehyde - 3 Phosphate 1.3 - Diphosphoglyceric acid

3. 1,3 - Diphosphoglyceric acid is transformed into 3 phospoglyceric acid when one of the phosphate groups is released, which is accepted by ADP to synthesize a molecule of ATP. *Phosphoglyceric kinase* enzyme catalyzes this reaction.

The enzyme has Mg^{++} as the cofactor

1.3 Diphosphoglyceric acid + ADP 3 Phosphoglyceric acid + ATP

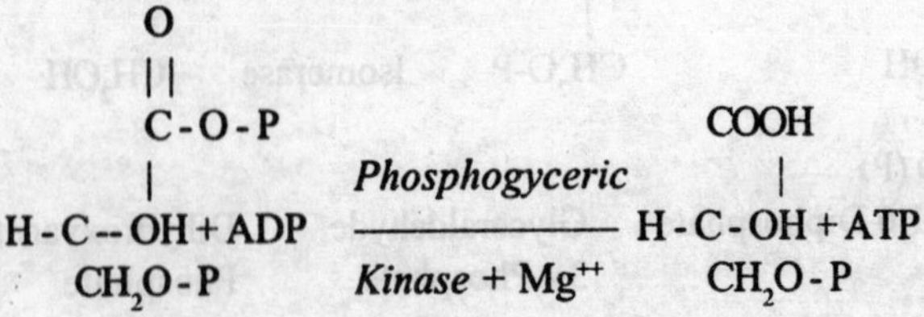

4. 3 Phosphoglyceric acid gets transformed into 2 phosphologlyceric acid (phosphate group is shifted from the 3rd carbon position to the 2nd carbon position), under the influence of the enzyme *Phosphoglyceromutase.*

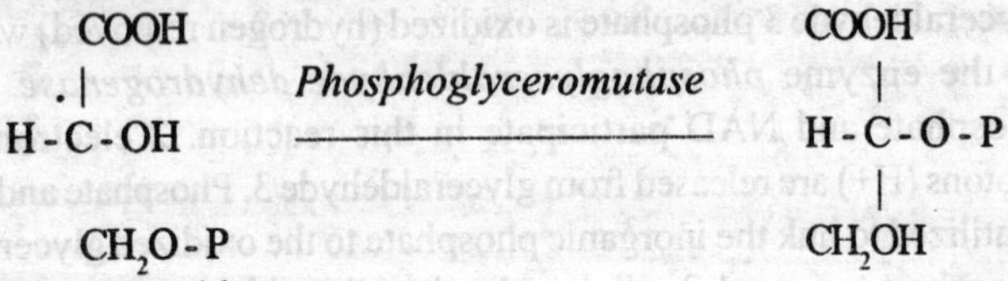

3. Phosphoglyceric acid

5. 2 - Phosphoglyceric acid (2PGA) gets converted into phosphoenolpyruvic acid (PEP) under the influence of the enzyme, *enolase* with Mg^{++} ions as cofactor. A molecule of water is eliminated from 2PGA in the process with the result an energy rich (-P) centre is created.

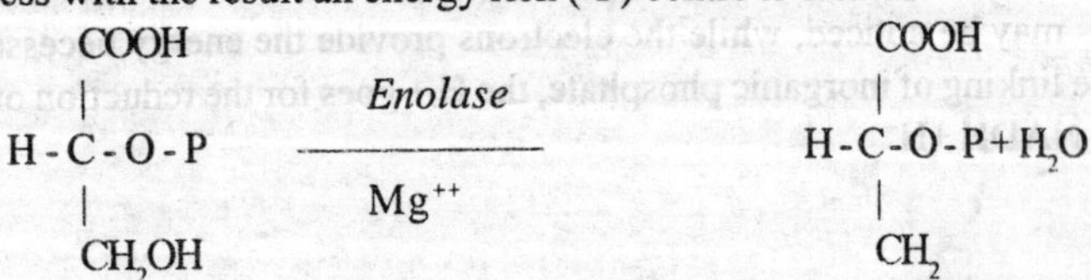

(2 PGA) (2 PEP)

6. In the molecule of PEP, much of the energy is centered round the phosphate group. This energy rich phosphate group is removed from PEP, which is accepted by ADP to produce a molecule of ATP. In the process, the dephosphorylated PEP becomes converted to pyruvic acid. This reaction is catalyzed by the enzyme *pyruvate kinase* with Mg++ and K+ ions as cofactors.

$$\begin{array}{c} COOH \\ | \\ H-C-O-P \\ | \\ CH_2 \end{array} + ADP \xrightarrow[Mg^{++}K^{++}]{\textit{pyruvate kinase}} \begin{array}{c} COOH \\ | \\ C=O \\ | \\ CH_3 \end{array} + ATP$$

(Pyruvic acid)

With the formation of pyruvic acid, the reactions of glycolysis come to an end. One molecule of pyruvic acid is produced for every triose, hence for hexose, it should be two molecules of pyruvate.

ATP account during glycolysis

The production of pyruvic acid terminates glycolysis. When we briefly glance through the glycolytic reactions we realize that there are two important stages – (i) phosphorylation and (ii) oxidation of Fructose 1.6 diphosphate to pyruvic acid.

During phosphorylation of glucose to form Fructose 1.6 diphosphate, two molecules of ATP are used up. Thus during phosphorylation of glucose, 2 molecules of ATP are used.

In the second stage, at two stages ATP molecules are synthesized–once during the conversion of 1,3 - diphosphoglyceric acid to 3 phosphoglyceric acid (2 molecules - i.3., one for every triose and two for every hexose), and again during the conversion of phosphoenol pyruvic acid to pyruvic acid (2 molecules of ATP, reactions are to be doubled while accounting for hexose). Thus in all, towards the end of glycolysis 4 molecules of ATP are produced and two molecules of ATP are consumed (during phosphorlation). Hence, the net gain of ATP during glycolysis is only two molecules.

But the statement that the net gain of ATP during glycolysis is only two molecules is not completely correct, because during the oxidation of glyceraldehyde 3 phosphate to phosphoglyceric acid, 2 molecules of NADH + H + are produced. These two molecules pass through the electron transport chain producing a total of 6 (3 + 3) ATP molecules (Details explained later). Hence, the net gain of ATP during aerobic glycolysis is, 6 + 2 = 8 molecules of ATP.

Breakdown of pyruvic acid

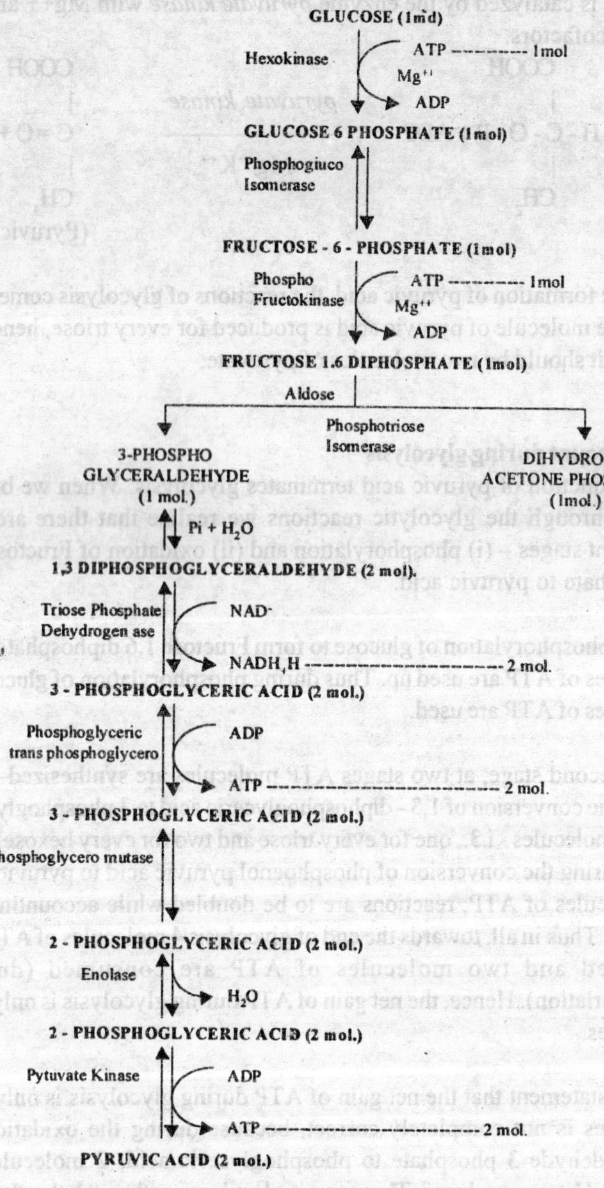

Fig. 6.2 Respiration

Glycolysis - Various reactions leading to the formation of Pyruvic acid

Further breakdown of pyruvic acid depends upon the nature of the organism, type of respiratory pathway and the presence or absence of O_2. Broadly these breakdown are –

(i) Alcoholic fermentation
(ii) Lactic acid fermentation
(iii) Amino acid synthesis and
(iv) Aerobic breakdown to produce CO_2 and H_2O

Options (i) and (ii) can be carried out in the abscence of O_2 and (iv) is possible in the presence of O_2, which happens to be the major pathway. Option (iii) leads to the biosynthesis of alanine by transaminase reaction with glutamate. This is not discussed here as it is not relevant to the respiratory pathway.

FERMENTATION

Fermentation is essentially anaerobic respiration found in microorganisms like fungi and bacteria. In fermentation, the substrate for oxidation is extracellular (not present within the cell), since the organisms are basically saprophytic.

Fermentation as a process to produce alcohol was known to human beings since time immemorial, eventhough its scientific understanding dates back to only the middle of the 19th century. Experiments conducted by ***Louis Pasteur*** established that fermentation is carried out with the participation of living cells.

Buchner in 1897 found out that the extract of yeast cells (which does not contain living cells) can bring about fermentation of sugar solution. Subsequently, it was discovered that the enzyme *Zymase* present in the yeast extract is capable of fermenting activity (breakdown of carbohydrates together with the formation of alcohol and CO_2).

The enzyme *Zymase,* will be active only in the presence of phosphate. *Sucrase* and *Maltase* are the other carbohydrate fermenting enzymes secreted by yeasts.

In fermentation also, as in anaerobic respiration, atmospheric oxygen does not participate. Oxidation is accomplished by the intermolecular atomic shifts, with the result the product will have less energy than the original substrate. Much of the energy released during fermentation is wasted as heat, while the remainder is used for the conduct of metabolic activities.

$$CH_3COCOOH + NADH_2 \xrightarrow{\text{Dehydrogenase}} CH_3CHOH\ COOH + NAD$$

Pyruvic acid   Lactic acid

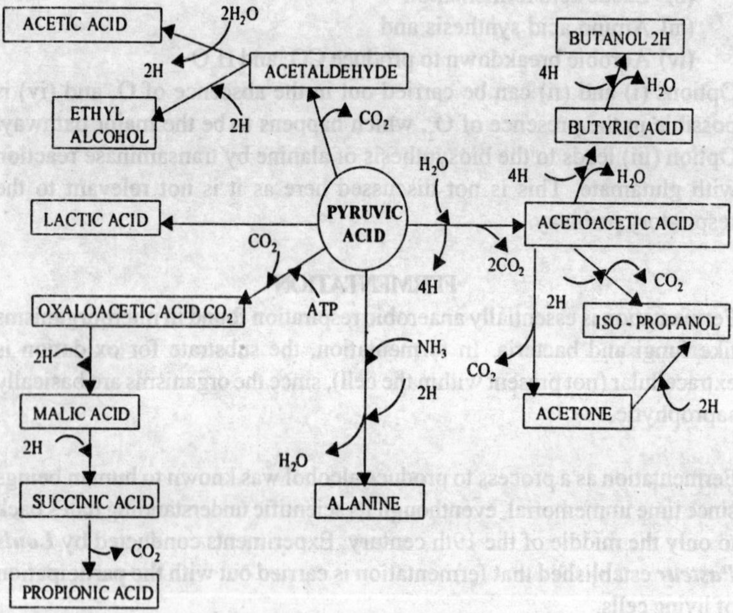

Fig. 6.3 Respiration
Anaerobic degradation of Pyruvic acid

The end product of fermentation need not always be alcohol, it may be various types of organic acids such as butyric acid, lactic acid etc. Depending on the end product, fermentation may be termed as alcoholic fermentation, lactic acid fermentation etc. The following are formulae for these fermentations.

(i) $C_6H_{12}O_6 \rightarrow 2C_2H_5OH + 2CO_2$ (ethyl alcohol)

(ii) $C_6H_{12}O_6 \rightarrow C_4H_8O_2 + 2H_2 + 2CO_2$ (butyric acid)

(iii) $C_6H_{12}O_6 \rightarrow 2_3H_6O_3$ (Lactic acid)

(iv) $C_2H_5OH + O_2 \rightarrow CH_3\ COOH + H_2O$ + energy (acetic acid)

In the formula (iv) given above, for the formation of acetic acid from ethyl alcohol, O_2 is necessary. Hence acetic acid fermentation is different from other types of fermentation.

Fermentation is carried out by microorganisms, a point which has been emphasized earlier. The following is a list of organism which carry out the various types of acid fermentation.

	Organism	Type of fermentation (based on end product)
1.	*Bacillus butyricus and Clostridium butyricum*	Butyric acid
2.	*Bactyerium lactic acid*	Lactic acid
3.	*Acetobacter aceti*	Acetic acid

The brief description of fermentation given above may now be followed by certain details of a few specific types of fermentation and a mechanism of regeneration of ATP.

ATP Generation by fermentation

During the fermentation of the glucose by microorganisms about 1–4 molecules of ATP may be produced. While there are large numbers of end products of fermentation starting from glucose, it is surprising that there are only a few reactions concerned with energy conversions of substrate level phosphorylation. Of these, the three most important reactions are discussed below.

1.1m3 - Bisphosphoglycerate + ADP ——————— 3 - Phosphoglycerate + ATP

2. Phosphoglycerate + ADP ——————— Pyruvate + ATP
3. Acetyphosphateor Butyrylphosphate + ADP ———— Acetate or Butyrate + ATP

Most of the fermenting microorganisms use only reaction 1 and 2 above, and use pyruvate compounds synthesised from acetyl - Co - A as the necessary hydrogen acceptors. In the above, reaction 1 is catalysed by phosphoglycerate kinase and reaction 2 is catalysed by pyruvate kinase, during the fermentation of one molecule of glycose only two or a maximum of 4 moleculcs of ATP arc generated and the following products may be formed: lactate, ehtanol, acetone, butyrate, n-butanol, 2-propoanol, 2,3 - butanediol, caproate, acetate, carbon dioxide, and molecular hydrogen.

In reaction 3 given, above the enzyme involved is acetate kinase, and it offers the advantage of additional ATP regeneration. Acteyl phosphate is produced from acetyl - Co - A by phosphotransacetylase.

acetyl - Co - A + P1 ——— acetylphosphate + Co - A

Acetyl phosphate may also be produced from sugar phosphates such as fructose - 6 - phosphate via reactions catalysed by phospoketolase.

The ability of microbes to utilise acetate kinase depends on whether they can evolve molecular hydrogen. No hydrogen acceptors need to be synthesised for reducing equavalents that may be liberated as molecular hydrogen.

Fermentations may be basically classified into two types :

1. Alcoholic fermentation,
2. Acid fementation

1. Alcoholic fermentation

When respiratory substrates are sugars, during anaerobic respiration many of the microbes ferment them and produce ethanol (ethyl alcohol). The wide variety of microorganisms are involved in ethanolic fermentation. The chief ethanolic producers are yeasts (particularly strains of *Saccharomyces cerevisiae)* and fungi. Yeasts also respire aerobically but in the absence of air, they ferment carbohydrates into ethanol and carbon dioxide. Among bacteria also a number of anaerobes or facultative aerobes produce ethanol as the main product during fermentation of sugars. Unusuall accumulation of ethanol can occur even in plants under anaerobic conditions.

Ethanol formation of yeasts

The fermentation process was infact studied in great detail in yeasts. The conversion of glucose to ethanol was formulated in 1815 by Guy – Lussac as follows.

$$C_6H_{12}O_6 —— 2CO_2 + 2C_2H_5OH$$

The normal fermentation of glucose by yeasts

The breakdown of glucose to ethanol and carbon dioxide by yeasts proceeds via the usual glycolytic pathway. From their onwards, the conversion of pyruvate to ethanol comprises few steps. In the first step, pyruvate is decarboxylated by pyruvate decarboxylase with the participation of thiamine pyrophosphate. As a result of this, a molecule of acetaldehyde is formed which is then reduced to ethanol by alcohol dehydrogenase with $NADH_2$ as the hydrogen donor. In this hydrogen transfer, the hydrogen obtained during the dehydrogenation of triphosphate is utilised. Thus the oxidation reduction balance in maintained.

History of yeast fermentation

Louis Pasteur was perhaps the first to have carried out experiments on fermentation by microorganims. It was Buchmer and Hahn (1896), however, who connected the fermenting activity with yeast. They observed that the extract obtained from Brewer's yeast by grinding with kieselguhr (diatomaceous earth), and sand produced gas bubbles after the addition of sugar. This was perhaps a first example of biochemical reaction *in vitro*. Harden and Young (1906), made the discovery that inorganic phosphate is a requirement for the fermentation of glucose by yeast extract and it is incorporated in fructose - 1,6 - bisphosphate. According to Harden and Young equation, fermentation of glucose by yeast is as follows.

$$2C_6H_{12}O_6 + 2Pi \longrightarrow 2CO_2 + 2C_2H_5OH + H_2O$$

+ fructose - 1,6 - bisphosphate

Fermentation formulae by Neuberg

The studies of Neuberg on fermentation are of great significance. He showed that yeast can ferment both glucose as well as pyruvate. The formation of intermediate compounds such as acetyldehyde could by demonstrated by trapping it with hydrogen sulphite. On addition of hydrogen sulphite to glucose fermenting yeast, acetyldehyde is precipitated as the addition product.

$$CH_3CHO + NaHSO_4 \longrightarrow CH_2 - CHOH - SO_3Na$$

Under the above conditions, glycerol is formed as a new product while the yield of ethanol and carbon dioxide gets reduced. This kind of fermentation has been exploited industrially for the production of glycerol. By trapping the acetyldehyde (using hydrogen sulphite), it is not allowed to function as hydrogen acceptor. As a result, dihydroxycetone phosphate becomes the hydrogen acceptor and is reduced to glycerol - 3 - phosphate which is then dephospharylated glucose + hydrogen sulphite ——— glycerol + acetaldehyde - sulphite + CO_2

The above fermentation which is a modification of the normal ethanolic fermentation by yeasts is known as Neuberg's second fermentation. The principle of trapping a metabolite along the pathway has now become known as the trapping method and has been extensively used in biochemical research.

Addition of alkali to fermenting yeast extract, also leads to the formation of glycerol because acetyldehyde undergoes dismutation to ethanol and acetate, and hence cannot function as a hydrogen acceptor. This reaction is known as Neuberg third fermentation and is as follows:

$$2\ GLUCOSE + H_2O \text{———} ethanol + acetate + 2\ glycerol + 2CO_2$$

The normal ethanolic fermentation is referred to as Neuberg's first fermentation and the reaction is as follows.

$$C_6H_{12}O_6 \text{—————} 2C_2H_5OH + 2CO_2$$

Relationship of yeast to oxygen

Under anaerobic conditions, yeasts ferment very actively but they can hardly grow. This is because yeasts are aerobes. On admission of air, fermentation decreases in favour of aerobic respiration. As shown in Pasteur effect, fermentation can be completely supressed by vigrous aeration. Pasteur discovered this effect, more than 100 years ago during his investigations on wine making. Pasteur effect, however is not restricted only to yeast but has been found in all anaerobic cells. Aeration of yeasts supports growth, but leads to a decline in glucose consumption and in the reduction of ethanol and carbon dioxide. This observation appears to be sensible from the stand point of energy conversion and suggests an ingenious regulatory mechanism. One molecule of glucose anaerobically yields only two molecules of ATP in contrast to the aerobic yield of 38 molecules of ATP per glucose molecule. Thus, the cell adopts by regulation of substrate consumption to the energy gain obtainable under two conditions.

The Pasteur effect involves several regulatory mechanisms working tegether. One of these perhaps applies at the level of phosphorylation and can be explained by competition for ADP and P.

A second regulatory mechanism perhaps is also responsible for the Pasteur effect. The enzyme phosphofructokinase is allosterically inhibited by ATP.

Harden and Young's explanation of fermentation balance

According to Harden and Young, fructose, 1,6 biphosphate accumulates due to lack to ATP utilisation for energy requiring reactions in the cell - free system. This leads to an excess of ATP balance. The yeast extract does not have any phosphates activity so that ADP must be constantly regenerated by phospharylation of excess glucose or fructose - 6 phosphate.

Industrial uses of yeast

For industrial purposes, different strains of yeast are selected. These are of two types.

Bottom yeast and *Top yeast.* Bottom yeasts are weekly respiring and actively, fermenting ones, and hence are used in the brewing of beer. Top yeasts on the other hand are used for production of ethanol and wine making.

Baker's yeast (*Saccharomyces cerevisiae),* are mostly top yeasts and are mainly used in baker's to produce carbon dioxide to raise the dough, i.e, these are grown in tanks with good aeration. In this reaction, ethanol is produced as a byproduct.

Brewer's yeasts are mostly bottom yeasts and are mainly used in brewing of beer from barley.

Industrial alcohol is produced from the residue of cane sugar refining called molasses. Occassionally, potatoes also may be used as rawmaterial for fermentation. Alcohol may also be obtained by hydrolysed wood chips. Microbial species used for these fermentations are species of *Candida* and *Endomyces.*

Ethanol formation by bacteria

Not many bacteria have been studied and exploited for ethanol production as has been done in the case of yeast. Among all the bacteria examined so far, only in the case of *Sacrcina ventriculi,* the pathway of ethanol formation is similar to that of yeast. Another bacterium that seems to produce ethanol has been isolated from the juice of the Agave plant. This bacterium (*Zymomonas mobilist*) breaksdown glucose via 2-keto-3-deoxy-6-phosphogluconate pathway and splits the pyruvate to acetyldehyde and carbondioxide under the influence of the enzyme pyruvate decarboxylase. The acetyldehyde is further reduced to ethanol along with carbondioxide and some quantity of lactic acid. It has further been discovered that the agave alcohol contains carbon atoms 2,3,5 and 6 of glucose, whereas a yeast ethanol contains carbon atoms 1,2,5 and 6 of glucose.

Some of the bacteria belonging to Enterobacteriaceae and Clostridia are also known to produce ethanol as a product of secondary fermentation. In these microorganisms, however, the precurosor of ethanol namely acetyldehyde is not produced from pyruvate by pyruvate decarboxylase but originates by the reduction of acetal Co-A.

Some of the acid fermenting bacteria like *Leuconostoc mesenteriodes,* produce alcohol by an entirely different pathway. Glucose is first broken down by the early steps in the pentose phosphates cycle to produce pentose phosphate. This pentose phosphate (xylulose -5 - phosphate) is then acted upon by the enzyme phosphoketolase to produce acteyl phosphate. This is reduced to ethanol by the enzymes acetaldehyde dehydrogenase and alcoholdehydrogenase. The remaining product of breakdown glyceroldehyde - 3 - phosphate is converted into pyruvate xylulose - 5 - phosphate + Pi — acetylphosphate + glyceraldehyde -3 - phosphate.

Acid fermentation

Anaerobic breakdown of sugars by certain microorganisms results in the formation of various types of acids such as lactic acid, acetic acid, butyric acid, propionic acid, formic acid etc. Some of these fermentation are discussed below in some detail.

Lactic acid fermentation

The lactic acid bacteria belong to the family *lactobacteriaceae* and includes a wide variety of rods and cocci. All the members of lactobacteriaceae whether it be the physical form have a characteristic physiology, are gram positive and non-sporing (with few exceptions). They are usually non-motile and have carbohydrate as the main energy source and produce lactic acid. Another family of bacteria which produce lactic acid is Enterobacteriaceae. The lactobacteriaceae, however, are obligate fermenters. Basically, lactobacteria are anaerobic but aerotolerant.

Generally, the members of lactobacteriaceae require many accessory factors in the culture medium in addition to the basic nutrients. Addition of vitamins, amino acids, purines and pyrimidines is necessary to cultivate lactobacteria. The culture medium is usually complex one containing yeast extract, tomato extract or even blood.

Generally, the metabolic machinery of lactobacteria appears to be quite defective as it cannot produce a number of compounds which are required for its growth and consequently they have to be added to the medium. Many microbiologists regard the lactobacteriaceae as "metabolic cripples", as they have lost the ability to synthesise a number of metabolants. This may be due to the fact that in the course of evolution they seem to have acquired an affinity to milk and other nutrient rich media. This has lead to a special metabolic mechanism in them to utilise lactose, an ability that most microorganisms do not possess. The ability to use lactose is also possessed by many intestinal bacteria of the coliform type.

Milk as such is produced as a nutrient factor by mammals and consists of lactose. It is not found in the plant kingdom. The ability of some microorganisms to utilise lactose may be regarded as an adaption to the environmental conditions prevailing in the mammalian digestive tract.

Lactose is a disaccharide which must be hydrolysed before it can enter the catabolic pathway to produce Hexoses. This reaction is catalysed by the enzyme β-galactosidase. This enzyme however, occurs only in, few bacteria. The product of hydrolysis mainly ngalactose is converted to glucose - 1 - phosphate after phosphorylation.

Occurrence of lactobacteria

Lactic acid bacteria are largely distributed in places where there is a generous amount of lactose, as such, they are hardly found in soil or water. The natural habitats of natural bacteria (Schlegel, 1993) are –

1. Milk and the places where milk is produced and processed (*Lactobacillus lactis, L. bulgaricus, L.Helveticus, L.casei, L.fermenttium, L.brevis, Lactococcus lactis, L.diacetilactis*).

2. Intact and rotting plants (*Lactobacillus plantarum, L.delbruackii, L.fermentum, L.brevis, Lactococcus lactis, Leuoconostoc mesenteriodes*)

3. Intestinal tracts and mucous membranes of animals and humans (*Lactobacillus acidophilus, Bifidobacterium, Enterococcus facealis, Streptococcus salivarius, S.bovis, S.oyogenes, S.pneumoniae*).

Classification of lactobacteria

They are classified into two basic categories based on their ability to ferment glucose into solely lactate or to other additional products. If the fermentation produce is only lactat the bacteria are said to be **homofermentative** and if the fermentation product is many other compounds in addition to lactate they are said to be **heterofermentative.** The following table gives a list of heterofermenting and homofermenting bacteria.

Cocci	**Rods**
Homodermentative : $C_6H_{12}O_6$ —— $2CH_3$ - CHOH - COOH	
Streptococci	Lactobacilli
Lactococcus lactis subsp. Lactis	Thermobacteria (temp. Opt 40°C,do not grow at 15°C)
Lactococcus lactis subsp. Lactis	subsp, Delbrueckii.
var.diacetilactis	Lactobacillus Delbrueckii
subsp.cremoris	Subsp. Lactis
Enterococcus faecalis	Lactobacillus delbrueckii
Streptococcus salivarius	subsp.bulgaricus
subsp.salivarius	Lactobacillus helveticus
Streptococcus salivarius	Lactobacillus acidophilus

subsp.thermophilus	Lactobacillus salivarius
Streptococcus pyogenes	Streptobacteria (temp.opt. 30-37°C, alwaysgrow at 15°C)
	Lactobacillus casei
	Lactobacillus alimentaris
	Lactobacillus coryniformis
	Lactobacillus plantarum

Heterofermentative : $C_6H_{12}O_6$ — CH_3 - CHOH - COOH + CH_3 - CH_2OH + CO_2 (or CH3 - COOH)

Streptococci	Lactobacilli
Leuconostoc mesenteriodes subsp. mesenteriodes	Lactobacillus bifermentans
	Lactobacillus brevis
Leuconostoc mesenteriodes subsp.dextranicum	Lactobacillus fermentum
	Lactobacillus kandleri
Leuconostoc mesenteriodes subsp.cremoris	Lactobacillus viredescens
Leuconosoc lactis	

Homolactic fermentation

Homofermentative lactobacteria produce as the end product of metabolism almost pure lactate. They breakdown glucose via the fructose biphosphate pathway as they posses all the required enzymes. Aldolase, which is a key enzyme helps in the anaerobic breakdown of glucose to lactate. From glucose pyruvate is produced first, which is then reduced to lactate. The hydrogen necessary for reduction of pyruvate comes from the dehydrogenation of glyceraldehyde - 3 - phosphate. The type of lactate formed whether it is D, L or DL depends on the stereospecificity of the enzyme lactate dehydrogenase. Only the small fraction of the pyruvate is decarboxylated and converted to acetate, ethanol, carbon dioxide or possibly acetation. The extent of the formation of subsidiary products besides lactose depends on the availability of the oxygen.

Hetero lactic fermentation

The hetero fermentive bacteria lack some key enzymes that are necessary for the breakdown of fructose biphosphate. These enzymes are aldolase and triosephosphate isomerase. The initial breakdown of glucose therefore is entirely along the pentose phosphate pathway i.e., via glucose - 6 - phosphate —— 6 - phosphoclugonate —— ribulose - 5 - phosphate. Ribulose - 5 - phosphate is further converted into xylulose - 5 - phosphate by the enzyme epimerase. This pentose sugar is cleaved by a TPP dependent reaction catalysed by the enzyme phosophoketolase resulting in the formation of glyceroldehyde - 3 phosphate and acetyl phosphate. Lactobacteria such as *Leuconostoc mesenteroides* ferment glucose to yield lactate, ethanol and carbondioxide. Obviously, they reduce the acetyl phosphate via acetyl - Co - A and acetyldehyde to ethanol. Many other heterofermentive bacteria convert the acetyphosphate either partially or completely to acetate, transfering the energy rich phosphate bond to ADP and thus producing ATP. The excess of hydrogen in this instance is transferred to glucose to give mannitol. The other compound glyceraldehyde - 3- phosphate is converted to lactate via pyruvate.

Fructose is also fermented by heterofermentive bacteria resulting in the formation of lactate, acetate, carbon dioxide and mannitol.

$$3 \text{ fructose} — \text{lactate} + \text{acetate} + CO_2 + 2 \text{ mannitol}$$

Another bacterium, *Lactobacillus plantarum* can ferment glucose homofermentatively, but breaks pentoses by the enzyme phosphoketalase to lactate and acetate. There are some bacteria such as *Lactobacillus casei,* which are homofermentive on glucose; become heterofermentive when the source substrate is ribose, and produce acetate and lactate.

Another important example of heterofermentive lactic acid bacteria is *Bifidobacterium bifidium,* which owes its name to the shape of its cells. This is predominantly present in the intestinal tract of breast fed babies. These bacteria depend heavily on sugars that contain N - acetylglucosamine, which is found only in human milk and not in any other milk. The species of *Bifidobacterium,* are strict anaerobes and require an atmosphere containing atleast 10% carbon dioxide. The metabolism of glucose by *Bifidobacterium* is as follows.

Glucose fermented via the phosphoketolase pathway. The microbe lacks aldolase and glucose - 6 - phosphagte dehydroganse, but has phosphoketolase. This enzyme can split fructose phosphate and xylulose - 6 - phosphate to acetylphosphate and erythrose - 4 - phosphate or glyceraldehyde - 3 - phosphate respectively.

$$2C_6H_{12}O_6 — 2CH_3\text{-}CHOH\text{-}COOH + 3CH_3COOH$$

Propionic acid fermentation

The bacteria that breakdown glucose to propionic acid are called *Propionibacteria.* Generally, these bacteria are present in the intestine of ruminants such as cattle and sheep where they play an important part in the formation of fatty acids particularly propionic and acitic acids. These bacteria are responsible for the conversion of lactate to propionate. *Propionibacteria* rarely occur in milk nor are they found in soil or water. They can be isolated by enrichment cultures under anaerobic conditions in a lactate yeast extract medium inoculated with Swiss cheese.

The following is a list of propionic acid fermenting bacteria.

Propionibacterium freudenreichii
P.acidi - propionici
P.acnes
Veillonella alcalescens (Micrococcus lactilyticus)
Clostridium propionoicum
Selenomonas
Micromonospora

Character of Propionibacterium

Cells are gram positive, non-motile rods and also non-spore forming. Propioni bacteria are strict anaerobes and are intolerant to atmospheric oxygen. Because of its intolerance, they are long regarded as obligatory fermentors. Some recent cytochromes and catalase. This suggests that under conditions of regulated aeration the cell growth perhaps will be better than under anaerobic conditions.

Hence propionibacteria are regarded as micro aerotolerant.

Under anaerobic conditions, propionibacteria ferment a variety of sugars such as glucose, sucrose, lactose and pentoses to porpionic acid. Even glycerol can be converted to propionic acid.

Wood and Werkman (1936), who have made a detailed study of glycerol fermentation by Propionibacterium acidi - propionici found out that the fermentation involves the net fixation of carbon dioxide. The fixed carbon dioxide was found in the excrete succinate. Carboxylation of pyruvate here is called, the Wood Werkman reaction.

$$3CH_3 - CHOH - COOH \longrightarrow 2CH_3 - CH_2 - COOH + CH_3 - COOH + CO_2 + H_2O$$

Reduction of lactate or pyruvate to propionate follows a pathway called methylmalonyl - Co - A pathway because of the characteristic intermediate compound methylmalonyl - Co - A. Pyruvate is carboxylated to oxalocetate by the enzyme methylmalonyl - Co - A. carboxytransferase. Oxalocetate is then reduced to succinate, this step is linked to an electron transport chain. The succinate gets converted to succinyl - Co - A and then gets transformed into methylmalonyl - Co - A by the enzyme metholmalonyl - Co - A mutase. This intermediate compound undergoes dcecarboxylation to yield propionyl - Co - A. Propionate is liberated from propional Co - A to succinate. Most of the propioni bacteria form propionate by the above mentioned pathway. The same pathway is used by Veillonella alcalescens and Selenomonas ruminatium also.

The methylmenloly - Co - A pathway is reversible and can be used to produce succinly - Co-.A.

Acryloly - Co - A pathway

Bacteria such as Clostridium propionicium, Bacteroides ruminicola and Megasphera elsedeni produce propionic acid by a simple route with derivatives such as acrylic acid, acryloyl - Co -A acting as the intermediate compound. This results in the formation of propionil - Co -A leads to the formation of propionate.

Formic acid fermentation

Some acid producing fermenting bacteria are collected together in a physiological grouping whose characteristic feature is the production of formic acid. In these bacteria, besides formic acid several other acids are also formed. Systematically, most of these bacteria are included in the family enterobacteriaceae. Formic acid producing interbacteria are gram negative, peritrichous, high motile and non spore forming rods. Physiologically they are facultatively aerobic and can obtain the energy either by aerobic, respiration or by fermentation.

Some of the common enterbacteriacease which produce formic acid are *Escherichia coli, Proteus, Enterobacter, Bifidiobacterium etc.*

Fermentation products

It has been pointed out early, that the fermic acid producing bacteria are also called mixed acid producing bacteria as they produce several acids in addition to formic acid. Many of the enterobacteriaceae bacillus species produces a number of organic products such as acetate, formate, succinate, lactate, ethanol, glycerol, acetoin etc. Sugars are largely metabolized via the fructose 1.6 diphosphate route and to a lesser extent by the pentose phosphate pathway.

Fermentation of glucose by Escherichia coli is characterized by the following reactions (Schlegél, 1986).

1. Conversion of pyruvate to acety Co - A and formate
2. Cleavage of formate to carbon dioxide and hydrogen
3. Reduction of acety Co - A to ethanol

The cleavage of pyruvate to acetyl Co - A and formate occurs only under anaerobic conditions and is catalysed by the enzyme pyruvate, formatelyase. This enzyme is very sensitive to oxygen. It is maintained in the reduced state by flavodoxin and requires for its activation, S - adenysyl - L - methionine.

Formate is split into carbon dioxide and hydrogen by most strains of C.coli as well as by other gas forming members of enterbacteriaceae. This reaction is catalyzed by the enzyme system formate: hydrogen lyase. The reaction is as follows :

$HCOOH \longrightarrow H_2 + CO_2$

The above reaction most probably involves the combined effects of a formate dehydrogenase and a hydrogenase

1. $HCOOH + X \longrightarrow CO_2 + XH_2$
2. $XH_2 \longrightarrow X + H_2$

Butyric acid - Butanol fermentation

Butyric acid, butanol, 2 - propanol and few other organic acids are produced as end products of carbohydrate fermentation by anaerobic bacteria belonging to the genus clostridium.

The genus clostridium belongs to the family Bacillaceae. The members are gram positive, motile (petritrichous), spore forming rods. Physiologically, clostridia are characterized by their intense fermentive metabolism. Species of clostridium, however, range from strict anaerobes (C.pasteurinum) to aerotolerants (C.acetobutylicum).

For their fermenting activity, clostridia use a variety of substrates such as polysaccharides, nuclei acids, proteins, purines, pyrimidnes etc.

Biochemistry of fermentation

Clostridal fermentation produces varying amounts of acids such as butryates, acetate, lactate etc. as well as alcohols (butanol, ethanol etc). In some instances, the end product of metabolism may even be acetone and gas (CO_2 or H_2.)

Clostridia metabolise glucose via the glycolytic pathway. Hydrogen removed from glyceraldehyde 3. Phosphate is added to organic acids synthesized from pyruvate and acetyl Co - A. Glucose fermentation by C. Butyrium and C. Acetobutylicum represents fermentation typical of the genus.

Butyrate is the product of condensation of two molecules of acetyl Co - A to acetoacetyl Co - A and subsequence reduction. Condensation of molecules is catalysed by the enzyme thiolase. Acetoacetyl Co - A is reduced by β-hydroxy butyryl - Co - A dehydrogenase and $NADH_2$ β-hydroxy butyryl Co - A. The enzyme corotonase removes water from this to give crotonyl CoA. This is then reduced by butyryl CoA dehydrogenase to butyryl CoA. The CoA can be transfered to acetate by CoA transferase to produce butyrate.

In a purely butyric acid fermentation, the hydrogen derived from the oxidation of pyruvate is liberated in gaseous form. When glucose is fermented to butyrate, CO_2 and H_2, a hydrogen balance is established and 3 mol ATP is regenerated per mol of glucose.

The end products of fermentation in Clostridium acetobutylicum are, butanol, butyrate, acetone and 2 - propanol. Initially, butyrate is excreted; with the increasing acidity, however acetoacelate decarboxylase enzyme becomes induced whose activity results in the formation of acetone and butanol. To be reduced to butano, butyrate must first get converted to butyryl CoA.

Fermentation of glutamic acid

Fermentation of glutamic acid attracted the attention of microbiologists, mainly because of the use of monosodium glutamate as a flavour increasing agent in food industry.

Japanese scientists devised a single stage fermentation to produce glutamic acid from strach (potato), using Micrococcus and Brevibacterium.

Presently, however, commercial production of glutamic acid is a two stage operation. In the first stage α- Ketoglutaric acid is produced by carbohydrate metabolism. A wide variety of bacteria produce α-Ketoglutaric acid (Pseudomonas fluroscens, Bacterium ketoglutarium etc). α Ketoglutaric acid is then converted to glutamic acid by the process of transamination. E.coli and B. Ketoglutaricum produce glutamic acid with alanine or aspartic acid acting as amino group donor. The reactions take place as follows :

α Ketoglutaric acid + Aspartic acid — glutamic acid + Oxalo acetic acid

Homoacetate fermentation

Bacterial activity that metabolizes substrate and ultimately proudces predominantly acetate is called homoacetate fermentation. Many of the species of Clostridia (C.formicoaceticum, C.thermoaceticum, C.acidicum etc), are able to transfer the hydrogen equivalents liberated in the initial oxidation of substrate only to carbon dioxide producing acetate.

$$8\,[H] + 2CO_2 \text{——} CH_3 \text{—} COOH + 2H_2O$$

C. Thermoaceticum and C.Formicoaceticum ferment glucose along the glycolytic pathway and produce almost 3 mol acetate per mol of glucose. Carbon dioxide released due to the decarboxylation of pyruvate gets refixed, and serves as H_2 acceptor. Pyruvate gives rise to acetate, CO_2, FdH_2 and ATP by the enzymes ferrodoxin oxidioreductase, phsophotransacetylase and acetate kinase.

MICROBIAL METABOLISM AEROBIC RESPIRATION

In the previous pages, anaerobic breakdown of carbohydrates has been studied ultimately leading to the formation of incompletely oxidised carbon compounds along with some amount of energy. In this chapter, we will study the aerobic breakdown of pyruvic acid leading to the complete oxidation of organic carbon and release of a large amount of energy. The essential feature of aerobic respiration is the participation of atmospheric oxygen.

Aerobic oxidation of pyruvic acid

This is the most important phase of aerobic respiration in which pyruvic acid gets converted into a two carbon compound called acetaldehyde, before entering the Krebs cycle for complete decarboxylation and oxidation. The reaction is mediated by a complex enzyme with atleast five essential cofactors (Gunsalnus 1954). The acetaldehyde combines with co enzyme A to form acetyl CoA which enters the Krebs cycle.

Formation of Acetyl Co A

As already pointed out, the enzyme which catalyzes the decarboxylation of pyruvic acids a complex one which includes 3 enzymes - *pyruvic acid decarboxylase, dihydroxylipoyl, transacetylase* and *dihydrolipoly dehydrogenase.* The enzyme complex also requires five cofactors – Thiamine pyrophosphate (TTP), Mg ions, NAD +, Coenzyme A (CoA) and lipoic acid.

The first step includes the formation of a complex between TTP and pyruvate, followed by the decarboxylation of the latter. In the second step, the acetaldehyde fragment reacts with the cofactor lipoic acid to form an *acetyl lipoic acid complex.* In this reaction, acetaldehyde is oxidized to the acid and lipoic acid is reduced. In the third step, acetyl group is removed from lipoic acid and added on to CoA, resulting in the formation of Acetyl CoA and reduced lipoic acid. In the final step, oxidized lipoic acid is regenerated by transforming the electrons from reduced lipoic acid to NAD +. This is necessary because, there should be a continuous supply of oxidized lipoic acid to accept pyruvate and produce acetyl CoA. The NAD which gets reduced to foom NADH + H^+ eventually enters into the electron transport system to produce 3 molecules of ATP. The following reactions summarize the formation of Acetyl CoA from pyruvate.

(i) Pyruvate + TPP $\xrightarrow{Mg^{++}}$ TPP Complex + Co_2

(ii) TPP Complex + Lipoic acid (oxidized) ——— Acetyl lipoic acid Complex + TPP

(iii) Acetyl lipoic acid complex + CoA ——— Acetyl lipoic acid (reduced form)

(iv) Lipoic acid (reduced form) + NAD ——— Lipoic acid (oxidized form) + NADH + H^+

The net result may be given by the following equation.

Pyruvate + CoA + NAD Acetyl CoA + CO_2 + NADH + H^+
Finally, the two carbon compound Acetyl CoA formed in cytoplasm enters the mitochondria to react with the acids of the 'Kreb's cycle for the removal of remaining 2 carbon and to produce energy to be stored in the molecules of ATP.

KREBS CYCLE

Acetyl CoA, the two carbon compound undergoes further decarboxylation when it enters a cycle of interconverting carboxylic acids present in mitochondria. Together with decarboxylation, there is oxidation also (dehydrogenation), resulting in the formation of molecules of ATP (to trap the released energy).

Krebs cycle (named after its discoverer), was actually worked out by Wood et al (1942), and Krebs (1943). The cycle has several names – Tricarboxylic XCycle (TCA), Citric acid Cycle, Mitochondrial respiration etc. Basically, the cycle consists of several 6 carbon, 5 carbon and 4 carbon acids. When these are getting interconverted, Acetyl CoA enters facilitating the interconversion, and in the process undergoes decarboxylation at two steps. The details of individual reactions are given below.

1. Formation of citric acid

Acetyl CoA, the two carbon compound reacts with a 4 carbon carboxylic acid - *oxaloacetic acid* in the presence of water to form a six carbon carboxylic acid, *citric acid.* This reaction is mediated by the enzyme *citrate synthetase.* In the process CoA is released which goes back to form some more Acetyl CoA.

$$\begin{array}{llll} CO_2COOH & & & CH_2COOH \\ | & & \textit{Citrate} & | \\ CH_2COOH + CH_3COS.CoA + H_2O & & \overline{} & C(OH)COOH + \\ & & \textit{Synthetase} & CoA + HS \\ & & & | \\ & & & CH_2COOH \\ \text{Oxalo acetic acid} & \text{Acetyl CoA} & & \text{Citric Acid} \end{array}$$

2. Citric acid loses a molecule of water to form cis *aconitic acid* under the influence of the enzyme *Aconitase* with Fe^{++} as cofactor.

$$\begin{array}{lcl} CH_2COOH & & CH_2COOH \\ | & \textit{Isocitric} & | \\ CH\,COOH + NAD & \overline{} & CH\,COOH + H_2O \\ | & Fe^{++} & | \\ CH_3COOH & & CH\,COOH \\ \text{Citric acid} & & \text{Cis aconitic acid} \end{array}$$

3. Cis aconitic acid now gets converted into *Isocitric acid* (isomer of citric acid) by the addition of a molecule of water. The enzyme *Aconitase* catalyzes this reaction also

$$\begin{array}{lcl} CH_2COOH & & CH_2COOH \\ | & \textit{Aconitase} & | \\ C\,COOH + H_2O & \overline{} & CH\,COOH \\ | & Fe^{++} & | \\ CH\,COOH & & CHOH\,COOH \\ \text{Cis aconitic acid} & & \text{Isocitric acid} \end{array}$$

The addition of water molecule to cis aconitic does not produce Citric acid, but its isomer isocitric because H_2O combines in a different position in the molecule of a cis aconitic acid.

4. Isocitric acid undergoes oxidation (removal of hydrogen), under the influence of the enzyme Isocitric dehydrogenase to form *oxalosuccinic acid.* The released H atoms are accepted by NAD to become NADH + H + The NADH + H + gets reoxidised after entering into the electron transport chain (see later).

$$\begin{array}{lcl} CH2COOH & & CH2COOH \\ | & \textit{Isocitric} & | \\ CH\,COOH + NAD & \overline{} & CH\,COOH + NADH + H + \\ | & \textit{dehydrogenase} & | \\ CHOH\,COOH & & CO\,COOH \\ \text{Isocitric acid} & & \text{oxalosuccinic acid} \end{array}$$

5. Oxalo succinic undergoes decarboxylation to form a 5 carbon acid, α *ketoglutaric acid.* The reaction is catalysed by the enzyme *decarboxylase* with Mn^{++} as the cofactor.

$$\begin{array}{l} CH_2COOH \\ | \\ CHCOOH \\ | \\ COCOOH \end{array} \xrightarrow[Mn^{++}]{\textit{Decarboxylase}} \begin{array}{l} CH_2COOH \\ | \\ CH_2COCOOH + CO_2 \end{array}$$

6. The next step is the oxidation of a α ketoglutaric acid which is achieved by the removal of H. Simultaneously, it also undergoes decarbosylation leading to the formation of Succinyl CoA. The enzymes involved are α *ketoglutaric dehydrogenase* and α *ketodecarboxylase,* together with coenzymes NAD and CoA. NAD gets reduced to NADH + H^+, while a molecule of CO_2 is liberated . Together with CoA, TPP, Mg^{++}, FAD and lipoic acid also participate in the reaction.

$$\begin{array}{l} CH_2COOH \\ | \\ CH_2COCOOH \\ \alpha \text{ Ketoglutaric acid} \end{array} + HS.Coa + NAD \xrightarrow[Mg^{++}\ TPP,\ \text{Lipoic acid, FAD}]{\textit{Dehydrogenase}} \begin{array}{l} CH_2COOH \\ | \\ CH_2COS\,CoA \\ \text{Succinyl CoA} \\ + NADH + H^+ + CO_2 \end{array}$$

7. Succinly CoA is hydrolysed with a molecule of H_2O to *succinic acid,* regenerating CoA for continued participation in the cycle. Inorganic phosphate and a phosphorylating enzyme are involved in the formation of succinic acid. The release of CoA accompanies the release of energy, which will be used to synthesize a molecule of Guanosine triphosphate (GTP) by the addition of energy rich phosphate to Guanosine diphosphate (GDP). This is the only place where GTP instead of ATP is synthesized. This type of phosphorylation is called **substrate level phosphorylation.** The molecule of GTP subsequently reacts enzymatically with ADP to produce ATP and GDP.

(i) Succinyl Co.A + ip Sukccinyl-P + Co.A
(ii) Sukccinyl 1-P + GDP Succinic acid - GTP
(iii) GTP + ADP GDP + ATP

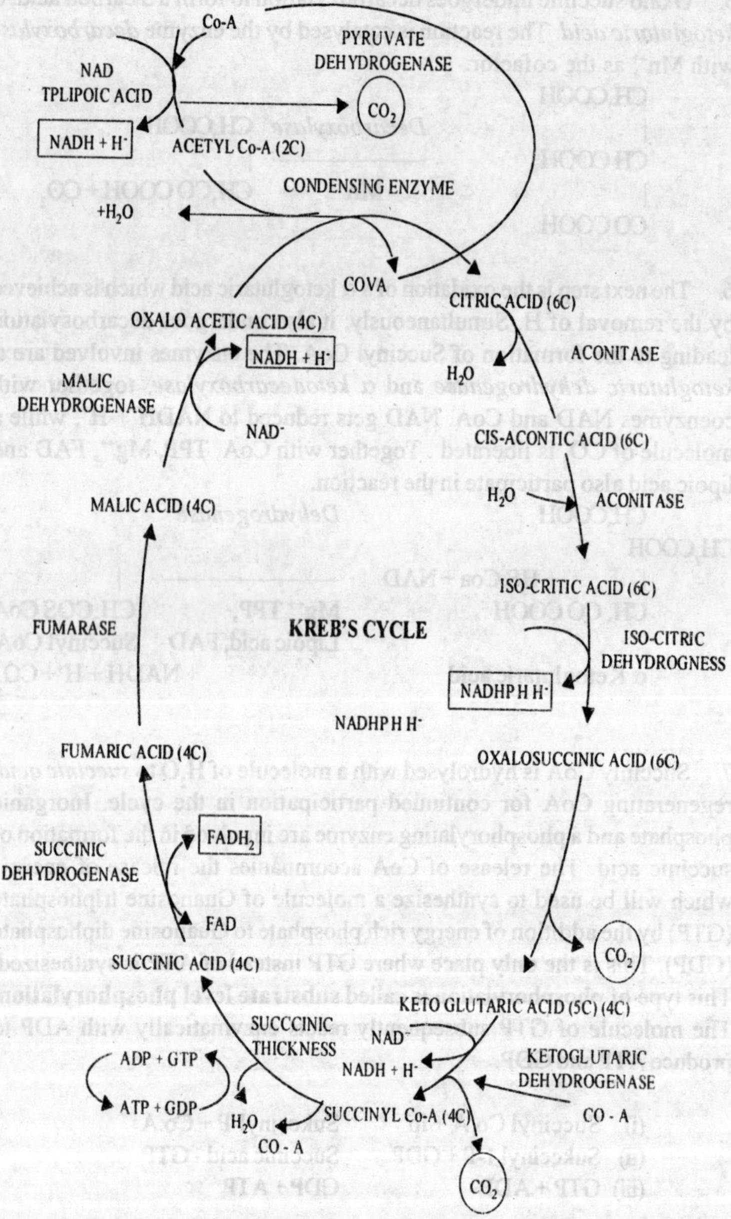

Fig. 6.4 Respiration
Reactions of Kreb's cycle

The enzyme succinate thiokinase catalyzes the reaction of conversion of succinyl CoA to succinic acid. From this reaction a molecule of ATP is synthesized.

CH_2COOH *Succinate* CH_2COOH
|+ GDP + Pi ———————— |
CH_2CO - S.CoA *Dehydrogenase* CH_2COOH
+ GTP + HS.CoA
Succinyl CoA Succinic acid.

8. Succinic acid undergoes dehydrogenation in the presence of a flavoprotein enzyme – succinic acid dehydrogenase, to form fumaric acid. The pair of hydrogen atoms released are accepted by FAD (this is the only reaction in Krebs cycle where hydrogen is not accepted by NAD) and it gets reduced to $FADH_2$.s

CH_2COOH *Succinic acid* CH COOH
| + FAD ———————— | + $FADH_2$
CH_2COOH *Dehydrogenase* CH COOH
Succinic acid Fumaric acid

9. One molecule of water is added to Fumaric acid enzymatically (Fumarase) to produce *malic acid.*

CH COOH *Fumarase* CHOH COOH
| + GDP + Pi ———————— |
CH COOH CH_2COOH
Fumaric acid Malic acid

10. In the last oxidation reaction of Krebs cycle malic acid undergoes dehydrogenation under the influence of *malate dehydrogenase* to prodeuce *oxalo acetic acid,* which combines with acetyl CoA to produce citric acid and continues the cycle. The hydrogen released from malic acid is accepted by NAD to produce NADH + H^+

CHOH COOH *Malate* CH COOH
| + GDP + Pi ———————— |
CH_2COOH *Dehydrogenase* CH_2COOH
Malic acid Oxaloacetic acid.

Summary of Krebs Cycle

In the Krebs cycle, there are four 6 carbon acids – *citric, cisaconitic, isocitric* and *oxalosuccinic* acids, one 5 carbon acid α *Ketoglutaric acid* and four 4 carbon acids *succinic, fumaric, malic* and *oxaloacetic* acids. When the 2 carbon acetyl CoA enters the cycle decarboxylation takes place at two stages – once to form the α Ketoglutaric acid and again to form succinic acid. With this, all the carbon atoms of the original glucose molecule have been removed.

The reactions of the Krebs cycle are for one pyruvate; since two molecules of pyruvate (3C) are formed at the end of glycolysis, the reactions of the Krebs cycle are to be doubled to account for complete glucose. The removal of CO_2 from glucose may be summarized as follows

Glycolysis

$CH_6H_{12}O$ ——— $2CH_3CO\,COOH$
Glucose   Pyruvic acid

In the above, there is no decarboxylation as all six carbon atoms are found in the organic molecule of pyruvate.

2CH3CO COOH   2CH3CHO + 2CO2
(Pyruvic acid)   (Acetaldehyde)

In the above, there is removal of two carbon atoms. Four more carbon atoms are left. They are removed as follows in the Krebs Cycle.

(i) 2 Molecules of Oxalosuccinic acid (6C) ——— 2 molecules of α Ketoglutaric acid + $2CO_2$ (5C)

(ii) 2 Molecules of α Ketoglutaric acid (4C) ——— 2 molecules of succinic acid (4C) + $2CO_2$

Thus in the aerobic respiration which involves Krebs Cycle, there is complete removal of all the organic carbon atoms to inorganic CO_2.

Another important aspect of Krebs Cycle is the dehydrogenation. The energy rich electrons released from the carboxylic acids at the following stages enter into the electron transport system (ETS) to produce molecules of ATP.

Stages of Dehydrogenation in the Kreb's Cycle

(i) Conversion of Isocitric acid to oxalosuccinic acid
(ii) Conversion of α Ketoglutaric acid to succinyl CoA
(iii) Conversion of succinic acid to Fumaric acid
(iv) Conversion of malic acid to oxaloacetic acid

These are the stages at which the energy is trapped for the synthesis of ATP molecules. Besides the above, during the conversion of succinyl CoA to succinic acid there is substrate level phosphorylation (not through the ETS). All these are explained below in detail.

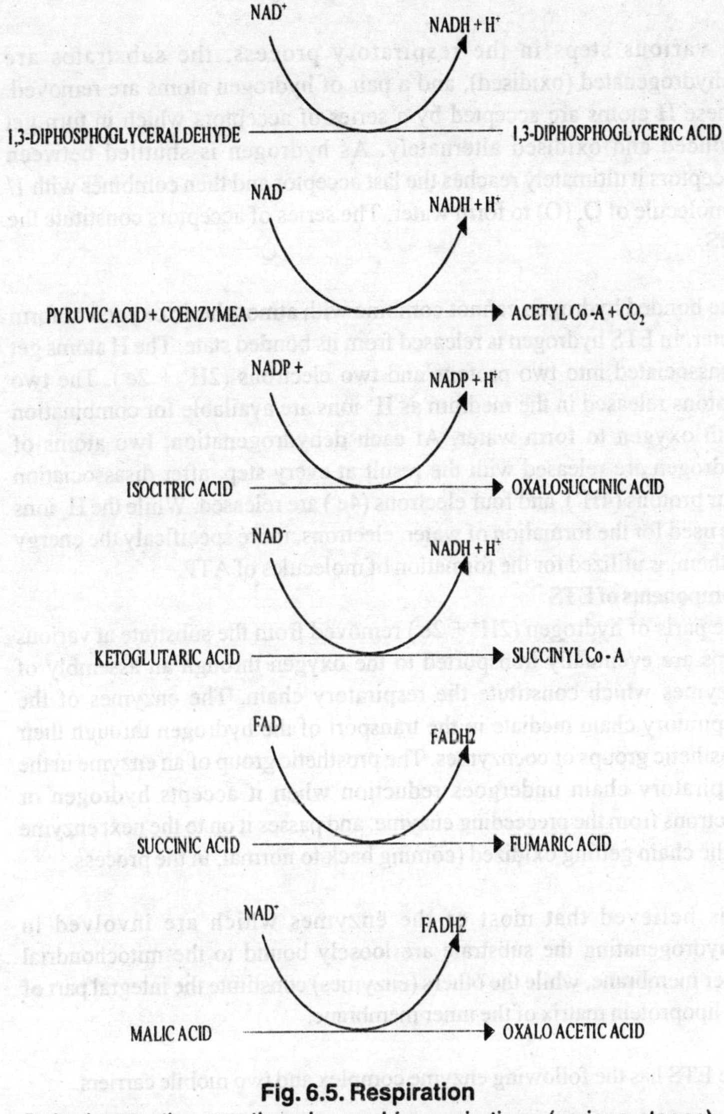

Fig. 6.5. Respiration

Dehydrogenation reactions in aerobic respirations (various stages)

ELECTRON TRANSPORT SYSTEM (ETS)

Electron transport system or oxidative phosphorylation is the final stage in aerobic respiration. Oxidative phosphorylation takes place in the F particle or oxysomes of the cristae in mitochondria.

In various steps in the respiratory process, the substrates are dehydrogenated (oxidised), and a pair of hydrogen atoms are removed. These H atoms are accepted by a series of acceptors which in turn get reduced and oxidised alternately. As hydrogen is shuttled between acceptors it ultimately reaches the last acceptor and then combines with 1/2 molecule of O_2 (O) to form water. The series of acceptors constitute the ETS.

The bonded hydrogen cannot combine with atmospheric oxygen to form water. In ETS hydrogen is released from its bonded state. The H atoms get disassociated into two protons and two electrons ($2H^+ + 2e^-$). The two protons released in the medium as H^+ ions are available for combination with oxygen to form water. At each dehydrogenation, two atoms of hydrogen are released with the result at every step, after disassociation four protons ($4H^+$), and four electrons ($4e^-$) are released. While the H_+ ions are used for the formation of water, electrons, more specificaly the energy in them, is utilized for the formation of molecules of ATP.

Components of ETS

The paris of hydrogen ($2H^+ + 2e^-$) removed from the substrate at various steps are eventually transported to the oxygen through an assembly of enzymes which constitute the respiratory chain. The enzymes of the respiratory chain mediate in the transport of the hydrogen through their prosthetic groups or coenzymes. The prosthetic group of an enzyme in the respiratory chain undergoes reduction when it accepts hydrogen or electrons from the preceeding enzyme, and passes it on to the next enzyme in the chain getting oxidized (coming back to normal, in the process.

It is believed that most of the enzymes which are involved in dehydrogenating the substrate are loosely bound to the mitochondrial inner membrane, while the others (enzymes) constitute the integral part of the lipoprotein matrix of the inner membrane.

The ETS has the following enzyme complex and two mobile carriers.

Complex I

This consists of a flavoprotein with attached $NADH_2$ dehydrogenase enzyme having FMN (Flavin mononucleotide) as the prosthetic group. 2–4 atoms of nonheme iron with iron-sulphur (Fe-S centre, is found in the enzyme.

Complex II

This also consists of a flavoprotein called *succinic dehydrogenase* with FAD (Flavin adenise (dinucleotide) as its prosthetic group. In addition, there are two Fe-S centres an i-e- S protein. Between Complex I and II, is present a mobile carrier ubiquinone (CoQ). CoQ is reduced by $NADH_2$ as well as succinic hydrogenase.

Complex III

This consists of Cyhtochrome *B* and C_1 with a non heme iron protein with only one Fe-S centre. However, the presence of Cytochrome C_1, is doubtful in plant ETS.

Complex IV

This is also called cytochrome oxidase or terminal oxidase. The complex consists of cytochromes *a* and a_3, having one or two atoms of firmly attached copper. Another mobile carrier is found between Complex III and IV which has a prosthetic group made up of iron prophyrin IX. This is identical with the heme of hemoglobin.

Complex V

This is the enzyme system called ATP ase and is mainly concerned with the synthesis of ATP from ADP and inorganic phosphate.

The plant electron transport system consists of these complexes and their location and sequence may be decided by the evidence provided by redox potential and respiratory inhibitors.

Evidence from Redox potential

In the biological system, certain compounds more readily get oxidized (give out electrons), while others get reduced (accept electrons). This phenomenon of release or acceptance of electrons is called **Redox potential** (E'ϕ). It is usually defined as the ratio between the reduced and oxidized forms of a compound. The standard electrode potential (Eo) may be defined as the potential at 50% oxidation for an oxidation reduction system at pH7. The standard electrode potential measurement (measured in milivolts) of the various components of the ETS gives the following sequence (in the order of decreasing potential), with exception of CoQ. The sequence is

NAD, FMN, FAD, *Cytb, Cytc, Cyte, Cyta,* and *Cyta3.* The standard redox potential of NADH2 is 0.32 volts while that of water is + 0.82 volts. It is for this reason that electrons from hydrogen to oxygen will yield energy.

Evidence from respiratory inhibitors

The correct sequence of the components of ETS can also be ascertained by the use of specific inhibitors, which block the function of one or the other components. For instance, when dilute cyanides are used which block the functioning of *Cyta* and *a*3, all other components get reduced i.e., function normally indicating *Cyta* and *a*3 are at the end. If, on the other hand antimycin A is used, it inhibits the functioning of NAD, all other components remain in the oxidized state (will not get reduced), thus indicating that NAD must be the first electron acceptor.

Stages of dehydrogenation

Total oxidation of the substrate material (glucose) into CO_2 and water involves several dehydrogenations at various steps (in glycolysis and Krebs cycle). The released hydrogen is picked up by the NAD or FAD and passed on to the next electron acceptor.

1. 2H is removed in glycolysis when a molecule of glyceraldehyde 3. Phosphate is converted into 1.3 Diphosphoglyceric acid. NAD picks up the hydrogen and gets reduced to $NADH_2$.
2. During the decarboxylation and dehydrogenation of pyruvic acid into acetyl CoA. 2H is removed. Hydrogen is accepted by NAD to become NADH2.
3. 2H are removed from isocitric acid to form oxalosuccinic acid. NAD accepts hydrogen and is reduced to $NADH_2$.
4. During the oxidation of α Ketoglutaric acid to succinic and 2H are removed. This is again accepted by NAD to become $NADH_2$.
5. During the formation of fumaric acid from succinic acid, dehydrogenation occurs which reduces FAD to $FADH_2$ (Note : This is the only instance where FAD, and not NAD is the first electron acceptor from the substrate).
6. During the dehydrogenation of malic acid to oxaloacetic acid, 2H atoms are removed which are picked up by NAD to get reduced to $NADH_2$.

Thus in the above 6 steps (one in glycolysis, one in pyruvate and four in Krebs cycle), a total of 12 hydrogen atoms are removed. Since this number is for triose only, it has to be doubled for a hexose and it gives a number of 24 hydrogen atoms. These hydrogen atoms travel down their electrical gradient along the various electron acceptors, which alternately get reduced and oxidized as the electron in shuttled through. At specific stages, the energy in the electron in made use of for the synthesis of ATP (oxidative phosphorylation).

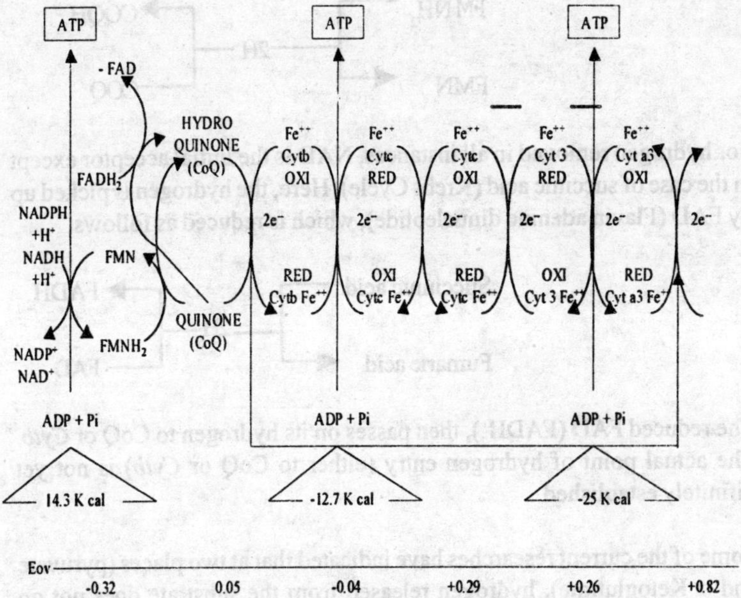

Fig. 6.6 Respiration
Schematic representation of electron transport system in oxidative phosphorylation

The Scheme of Electron Movement

The ETS begins with the acceptance of 2 atoms of hydrogen from the substrate by NAD which gets reduced to $NADH_2$ $NADH_2$ is oxidized when the 2H are transfered to the prosthetic group os lipoproteins of the FMN which in turn gets reduced to $FMNH_2$.

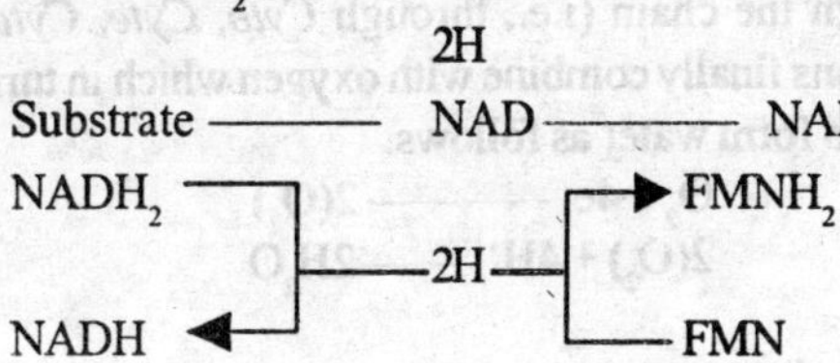

The 2H atoms are now transferred from $FMNH_2$ to the mobile carrier CoQ (Coenzyme Q or uniquinone). CoQ is chemically related to vitamin K and E and similar to plastoquinone of chloroplasts. This CoQ links the flavoproteins of the FAD to Cytochrome.

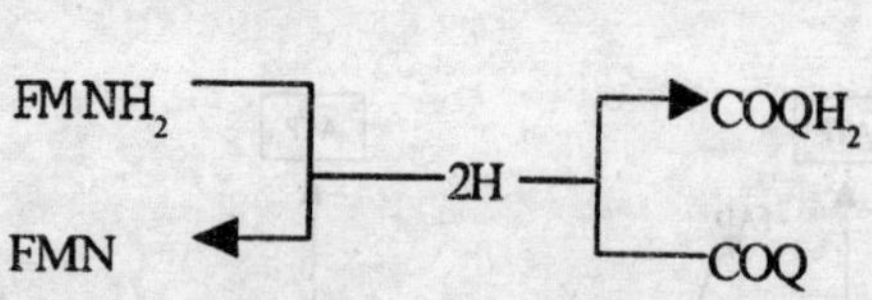

For hydrogen removed in all instances, NAD is the initial acceptor except in the case of succinic acid (Krebs Cycle). Here, the hydrogen is picked up by FAD (Flavin adenine dinucleotide), which is reduced as follows.

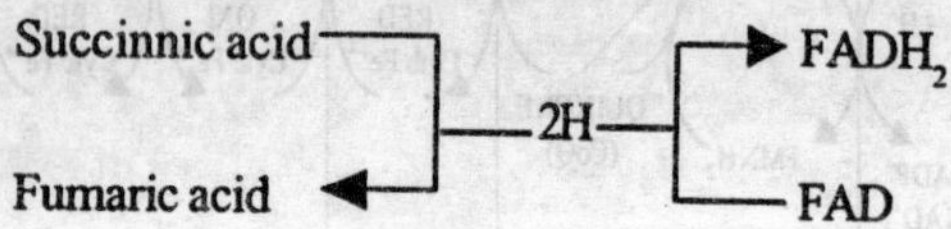

The reduced FAD ($FADH_2$), then passes on its hydrogen to CoQ or *Cytb*. The actual point of hydrogen entry (either to CoQ or *Cytb*) is not yet difinitely established.

Some of the current researches have indicated that at two places (pyruvate and α Ketoglutrate), hydrogen released from the substrate does not go direct to NAD but through lipoate and FAD and then to NAD. This is, however, subject to further proof.

According to Kligenberg and Kroger (1967), CoQ not only picks up hydrogen from succinate *via* FAD but also from α glycerophosphate and Acetyl CoA. The FAD of acetyl CoA, an additional flavoprotein ETF (electron transporting flavoprotein) is present.

The reduced CoQ is oxidized by the released of 2 hydrogen atoms. Soon after release, hydrogen undergoes disassociation into two, protons (2 H^+) are released into cytoplasm, while only the electrons are further transported through the chain (i.e., through *Cytb, Cyte, Cyta* and *Cyta* 3.) The two electrons finally combine with oxygen which in turn combines with protons (H+) to form water as follows.

$$O_2 + 4e^- \longrightarrow 2(O_2)$$
$$2(O_2) + 4H^+ \longrightarrow 2H_2O$$

(In the above equation, $4e^-$ and $4H^+$ are shown instead of $2H^+$ and $2e^-$. We have to remember that the reaction sequences of glycolysis (from 3 PGA onwards), and Krebs Cycle are for only half of glucose (one triose), hence they have to be doubled while accounting for a complete glucose molecule).

The passage of electrons are two at a time from NAD upto CoQ. But when electrons are transported from CoQ to *Cytb,* each molecule of *Cytb* can accept only one electron. But if we account for the passage of only one electron at a time, it will not be sufficient to reduce molecular oxygen. How the two electron chain is changed to a single electron chain (at the same time delivering two electrons towards the end), is as yet completely not known. The following are some of the views.

(a) Some unknown intermediate compound transfers the electrons one by one into cytochromes.
(b) The electron chain is forked at the point of CoQ so that two electrons can pass through the two branches (one in each branch). Each branch has the same enzyme system as the other. Finally, the two electrons released from the branches combine with molecular oxygen.

(c) The electron chain is formed at the point of CoQ so that the two electrons can pass through the two branches (one in each branch). Each branch has the same enzyme system as the other. Finally, the two electrons released from the branches combine with molecular oxygen.

The third view seems to be possible as it satisfactorily accounts for the smooth flow of both the electrons in separate but parallel channels.

The electrons give out by CoQ are first accepted by Cytochrome *b* which gets reduced. *Cytb* is reoxidised by transferring its electron to *Cyt c*. From these, the electron passes to *Cyt c, Cyt a,* and finally to *Cyt a*3. *Cyt a*3 gets oxidized by transferring electron to the atmospheric oxygen. How is it released? Each enzyme consists of a heme prosthetic group called *protoheme.* This iron alternately gets reduced and oxidized by accepting and donating an electron at a time as follows.

Cyt b CFe^{2+} Cyt*c* (Fe^{2+})

e

Cyt b (Fe^{3+}) Cytc_1 (Fe^{3+})

Some physiologists argue that *Cyt c*l is not a component of plant ETS. Hence, they believe that *Cyt b* transfers the electrons to *Cytc.*

ATP synthesis (oxidative phosphorylation)

All biological reactions require energy and this depends on the formation and utilization of, energy rich' phosphate bonds. A high energy bond does not actually mean that the bond is strong, on the contrary the bond is labile and when it is broken in a reaction, a large amount of energy is released. The energy requiring reactions always obtain energy by using certain compounds which readily release energy. These are called energy currencies. Similarly, during the storing of energy that is released in a reaction, these intermediate molecules are synthesized. Several molecules which act as energy intermediaries are known in biological systems. Of these, the most important are ATP (Adenosine Triphosphate), and ADP (Adenosine Diphosphate). Both ATP and ADP are derived from AMP (Adenosine Monophosphate). AMP consists of a purine (Adenine), a pentose sug (ribose), and phosphoric acid. ADP and ATP are similar to AMP except in the number of phosphate groups. While ADP has two phosphate groups, ATP has three. The second and third phosphate bonds are energy rich in ATP. On hydrolysis of the phosphate bond, energy equivalent to 7600 cal per mol is released.

To sum up, whereever energy is released, ATP molecules are synthesized and whenever energy is required ATP molecules are utilized. This may be illustrated by the following equations.

Synthesis

$$\text{ADP} + \text{Pi} \xrightarrow{\text{energy}} \text{ATP}$$

(inorganic phosphate)

Utilization

$$\text{ATP} + \text{X} \longrightarrow \text{X-(P)} + \text{ADP}$$

(X = Substrate that requires energy)

ATP molecules are synthesized in living systems by two methods.

(i) Photophosphorylation and
(ii) Oxidative phosphorylation.

In photophosphorylation (seen in green plants), radiant energy is used to add Pi to ADP to produce ATP. In oxidative phosphorylation, ATP is synthesized using the energy of oxidation.

Oxidative phosphorylation is the most significant functions of ETS. The number of ATPs produced may be explained on the basis of P/O index (phosphorylation/Oxygen). P/O index is the ratio of ATP produced for each atom of oxygen used.

Observations of Lehninger point out that the P/O ratio in the case of NAD linked ETS is three *i.e.*, at three sites (between NAD and molecular oxygen), question as to why there are only three energy producing sites, the answer may be provided as follows.

When the electron passes through the chain from one acceptor to the other, at every step some energy is released. The energy so released wil be equivalent to the difference in energy level between the electron donor and acceptor. At certain steps, the amount of energy released is sufficient to bind Pi to ADP, and hence a molecule of ATP is synthesized. As per some calculations, it has been noted that 0.28 volts of energy is sufficient to synthesize a molecule of ATP from ADP and Pi, and in the ETS, whereever the energy difference between the donor and acceptor is 0.28 volts or more, a molecule of ATP is synthesized. As has already been pointed out, in substrate systems linked to NAD there are three sites, where a molecule of ATP is synthesized.

These are

(i) NADH — FAD
(ii) *cyt b* — *Cyt c* or *c*, and
(iii) *Cyt a* — $Cyta_3$

In the oxidation of succinnic acid which is linked to FAD directly, obviously there are only two sites for the synthesis of ATP, hence only two molecules of ATP are synthesized.

The scheme of ETS and the sites of ATP molecule synthesis proposed by Bidwell (1979), are slightly different, though he also agrees that in NAD linked substrate systems there are three sites of ATP synthesis and in FAD linked systems there are only two. But their places the CoQ (Ubiqinone) between *Cyt b, Cyt c* and not earlier to the cytochrome systems (between FAD and Cytochrome b). This poses a problem for the movement of electrons as ubiquione requires 2 H^+ atoms for reduction and not just electrons. But before the 2H reaches the Cytochrome b, it would have undergone disassociation with two protons ($2H^+$) getting released into cytoplasm. Hence, Bidwelo (1979), believes that once again (after ionization before they reach Cyt b) $2H^+$ and $2e^-$ unite to form H, and this reduces the ubiquinone. From ubiquinone when 2 H has to reach *Cyt c,* there is ionization, $2H^+$ is released into water while $2e^-$ continues in the ETS. Hence in this chain, there is ionization of hydrogen twice and not once as believed by other physiologists.

Mechanism of ATP formation

Same as in Phosophosphorylation.

Balance sheet of energy in Respiration

(Account of ATP molecules synthesized). The following is a total of ATP molecules synthesized during complete breakdown of glucose to CO_2 and H_2O

Glycolysis

Reaction Sequence	ATP Used	ATP Synthesized
1. Glucose to Glucose 6 Phosphate	1	-
2. Fructose 6 Phosphate to Fructose - 1,6 - Diphosphat	1	-
3. 1.3 Diphosphoglyceric acid to 3 Phosphogyceric acid (1 x 2)	-	2
4. Phosphoenol pyruvic acid to pyruvic acid (1 x 2)	-	2
5. Dehydrogenation of Glyceraldehyde - 3 Phosphate to 1.3 Diphosphoglyceric acid (through ETS with participation of NAD) 3 x 2	-	6
Total	2	10

Net gain of ATP during Glycolysis = 8 molecules.

(Note : Reactions 1 and 2 are for entire glucose molecule, while, 3,4 and 5 are only for a triose, hence they are doubled)

Pyruvic acid Oxidation

Reaction Sequence	ATP Used	ATP Synthesized
Conversion of pyruvic acid to Acetyl CoA; Dehydrogenation through ETS with initial acceptor being NAD (3 x 2)		6

Krebs Cycle

(There is no ATP loss in Krebs Cycle)

	ATP Synthesized
1. Isocitric acid to oxalosuccinic acid dehydrogenation through ETS with NAD participation (3 x 2)	6
2. Conversion of Ketoglutaric acid to succinyl	

CoA, ETS with NAD as the initial acceptor (3 x 2) 6

3. Conversion of succinyl CoA to succinic acid; substrate level phosphorylation 2

No ETS (1 x 2)

4. Conversion of succinic acid to Fumaric acid ETS beginning with FAD and not Nad. Hence only two sites of ATP synthesis (2 x 2) 4

5. Conversion of malic acid to Fumaric acid. ETSbeginning with NAD (3 x 2)

6

Total 24

Total ATP used during the entire aerobic respiration = 2

Total ATP produced during glycolysis, pyruvic acid oxidation and Krebs Cycle. 10 +

6 + 24 = 40

Net gain of ATP molecule for aerobic oxidation of one molecule of glucose 40 -

2 = 38

Difference between photophosphorylation and oxidative phosphorylation

No.	Character	Photo-Phosphorylation	Oxidative Phosphorylation
1.	Occurrence	Photosynthesis	Respiration
2.	Cell organelle involved	Chloroplast	Mitochondrion
3.	Occurrence within the organelle	Thylakoid	Inner membrane Memgranes of cristae
4.	Need for molecular oxygen	Not needed	Needed
5.	Source of energy	Light	Reduction-oxidation reactions
6.	Output of ATP molecules	Less	More
7.	Fate of ATP molecules	Used up for the Reduction of CO_2	Released into cytoplasm and available for all metabolic reactions.

Efficiency of Respiration

A molecule of glucose consists of a total amount of 6,86,000 (▲F) calories of energy in the form of chemical bond. During the complete oxidation of a glucose molecule 6,74,000 (673.6 Kcal) calories of energy is released. Out of this, only a part is trapped in the energy rich phosphate bonds of ATP molecules, while the rest is released as heat. Each ATP molecule has about 7,600 calories of energy. The total amount of energy trapped during the breakdown of glucose depends on the number of ATP molecules formed. In case of oxidative phosphorylation, the total number of ATP molecules produced will be 38. This means a total of 38 x 7600 x 100/6,86,000, or about 40% of the total energy will be negotiable, while the rest (60%) is wasted. In the alternate pathways of glucose breakdown (pentose phosphate shunt), a similar amount of energy will be made available in the form of molecules of ATP. This gives an efficiency of 40% for aerobic breakdown of glucose.

In anaerobic respiration, only two molecules of ATP are formed per glucose molecule. The efficiency will be

$$C_6H_{12}O_6 \text{——} 2C_3CH_4O_3 + 4H$$

(F = 52 Kcal)

= (2 x 7.6/52) x 100 = **29.2%**

The energy efficiency for alcoholic fermentation would be (here also only two ATP molecules are formed)

$$C_6H_{12}O_6 \text{——} 2C_2C_5OH + 2CO_2$$

(F = 52 Kcal)

= (2 x 7.6/52) x 100 = **29.2%**

OXIDATION OF INORGANIC COMPOUNDS

In chemolithotrophs, energy is obtained by oxidation of inorganic compounds. Chemolithotrophs are generally identified by their ability to grow in mineral media and to obtain their cellular carbon from carbon dioxide.

Chemolithotrophs may be classified into obligate chemolithotrophs and facultative chemolithotrophs. Obligate chemolithotrophs generally cannot utilise organic compounds as their carbon source. It was earlier believed that for these organisms organic compounds are toxic, but some of the recent studies have shown that organic compounds are not completely toxic eventhough they partially inhibit the growth of microorganisms. In facultative chemolithotrophs, the main source of cellular carbon is CO_2; they can use even organic compounds as the source of their CO_2 requirement.

Oxidation of inorganic compounds is the chief source of energy both for obligate and facultative chemolithotrophs. Inorganic substances like ammonia, nitrate, hydrogen, sulphur compounds, iron compounds etc. get oxidised and release to glucose. All chemolithotrophs can also be called autotrophs. In the biosphere, they are the chief non-photosynthetic primary producers. The following are some of the examples of oxidation of inorganic compounds.

1. Oxidation of nigtrogenous compounds

A. Ammonia Oxidation

This is seen in organisms such as *Nitrosomonas* and *Nitrosocystis*. Here, the electron donor is ammonium ion and the electron acceptor is oxygen. The oxidation of ammonium ion to nitrate provides the necessary energy for CO_2 fixation. During the process, water is a byproduct. The reaction takes place as follows :

$$NH_4 + 3/20_2 \text{ ——— } No_2 - H_2O + 2H^-$$

Two steps are involved in the oxidationo of ammonia to nitrate. In the first step, ammonia is converted into hydroxylamine involving the expenditure of energy. Between hydroxylamine and nitrate, there is atleast one intermediate compound. The electrons liberated by hydroxylamine pass through a series of electron acceptors such as falvin, cytochromes (b,ca) and produce ATP. An ATP dependent reverse electron flow helps in the reduction of NAD to NADH2. The reduction of NADH is apparently at the cost of ATP. The nature of the intermediate compound between hydroxylamine and nitrite is still uncertain.

The conversion of ammonia to nitrite is called nitrification. This process also involves two steps. In the first step, ammonia is converted into nitrate by bacteria such as *Nitrosomonas* and *Nitrosocystis*. In the second step, the nitrite is oxidised to nitrate by nitrite oxidising bacteria like *Nitrobacter*. Nitrification is of great importance in the maintenance of nitrogen balance in nature.

The conversion of ammonia to nitrate and the release of ATP takes place as shown in the following equation:

NH_4 + ——— NH_2OH ——— Intermediate NO_2

|

e

Flavin ——— Cyt b ——— Cyt c ——— Cyt a ——— O_2

|

ATP

B. Nitrate Oxidation

As has been mentioned above the bacterium involved in nitrite oxidation is *Nitrobacter*. This bacterium utilises nitrite as the source of electrons and oxygen as the acceptor of electron. Oxidation of nitrite to nitrate is achieved at one step as follows :

$NO_2^- + 1/2\,O_2 \quad NO_3^-$

It has been found that Nitrobacter can obtain energy for its growth from the oxidation of formate. Enzyme such as nitrite oxidase and formate dehydrogenase are found in the same respiratory particle.

Oxidation of sulfur compounds

Many of the chemolithotrophs oxidise/reduce sulphur compounds to obtain their energy. These chemolithotrophs include sulphur bacteria both green and purple. The inorganic compounds that get oxidised are sulphide, thiosulphate, thiocyanate and elemental sulphur. *Thiobacillus* can oxidase sulphur and sulphur compounds such as sulphides and sulphites.

Oxidation of sulphur (S^0)

Many of the species of *Thiobacillus,* such as T.*Thiooxidans,* can convert both sulphur and sulphite to sulphite to sulphate. *T.Thiooxidans. T.Concretivorous* and *Thioparus* can aerobically oxidise elemental sulphur to yield sulphate. In the oxidation of sulphur to sulphate, sulphate is the intermediate compound. Experiments conducted with cell-free extracts of *T.Thiooxidans* shows that the elemental sulphur first reacts with oxygen to produce sulphite. This reaction is cartalysed by the enzyme oxygenase and is dependent on glutathione. The reaction is as follows:

$$SO_8 + 12O_2 + 8H_2O \longrightarrow 8H_2SO_3$$

The sulphite that is produced as a result of the above reaction noenzymatically reacts with elemental sulphur to yield thiosulphate as per the following reaction:

$$H_2SO_3 + SO \longrightarrow H_2S_2O_3$$

Oxidation of Sulphide

Some Thiobacilli are known to use sulphide. During the oxidation of sulphide to sulphate 8 electrons are released. Green and purple sulphur bacteria can anaerobically oxidise reduced sulphur compounds like H_2S. Both sulphide and elemental sulphur are metabolised due to the interaction of the enzyme bound thiol or dithiol. Sulphite and elemental sulphur react with a sulphydryl group to yield an intermediate RSSH.

This intermediate is oxidised to sulphite by sulphide oxidise. There is no generation of ATP here as the reaction is not linked to cytochrome electron transport. At the next step sulphite is oxidised to sulphate by the enzyme sulphite oxidase. As this reaction is cytochrome linked it can generate ATP through ADP.

Oxidation of sulphite to sulphate

Many of the thiobacilli are known to bring about oxidation of sulphite to sulphate. This reaction takes place through an AMP-dependent APS - (adenosine phosphorulphonate) reductase pathway.

APS is a high energy compound and it is formed due to the interaction of sulphite and AMP (Adenosine Monophosphate). This reaction is catalysed by the enzyme APS reductase.

$$2SO_3 \text{——} + 2\,AMP \xrightarrow{\text{APS reductase}} 2\,APS + 4e^-$$

At the same time inorganic phosphate is esterified to ADP, and the reaction is catalysed by the enzyme ADP sulphurlyase.

$$2APS + 2Pi \xrightarrow{\text{ADP sulphurlyase}} 2ADP + SO_4$$

ATP is now generated from ADP by the action of the enzyme adenylate kinase.

$$2\,ADP \text{————} AMP + ATP$$

The sum of the above reactions can be studied as follows :

$$2SO_3 \text{——} + AMP + 2Pi \text{————} 2SO_4 + \text{——} ATP + 4e^-$$

Iron oxidation

Some chemolithotrophs such as *Thiobacillus, Ferroxidans* can oxidase ferrous to ferric iron as follows :

$$4\,Fe_2^{+} + 4H^{+} + O_2 \text{————} 4\,Fe_3^{+} + 2H_2O$$

As a result of the oxidation, electrons that are released will flow from ferrous compound to cyt c and cyt a, and ultimately react with moleculare oxygen. A molecule of ATP is formed between cyt a and O_2.

Oxidation of hydrogen

Hydrogen is produced during the anaerobic degradation of organic substances in soil. Many chemolithotrophic bacteria have the ability to utilise this hydrogen. In order to oxidis hydrogen, bacteria should possess enzymes such as hydrogenase. It has been seen that hydrogen oxidising bacteria have evolved independently and the enzyme hydrogenase occurs in at least three different forms –

1. Cannot reduce NAD and is inhibited by CO. This is the commenest one.
2. It is called the nitrogenase type and it reacts with ATP and Fd resulting in the release of molecular hydrogen.
3. It is called the *Hydrogenomonas type,* it reduces NAD and is inhibited by CO.

Oxidation of hydrogen is of the following type:

1. Oxidation reduction
2. Nitrate reduction
3. Sulphate reduction
4. Methanogenesis

1. Oxidation reduction

This is seen in the bacterium *Hydrogenomonas.* Oxidative phosphorylation linked to the reduction of oxygen by molecular hydrogen provides energy necessary for growth.

$$H_2 + 1/2\ O_2 \longrightarrow H_2O$$

The energy released is used for the reduction of carbon dioxide to cellular carbon. Oxidation of hydrogen provides energy through which cellular synthetic activities take place. Hydrogenase enzyme can split hydrogen as follows :

$$H_2 \longrightarrow 2H^+ + 2e^-$$

The electrons released have a low electrifcal potential, hence the carriers also should have a low redox potential.

In *Hydrogenomonas,* the electron transfer chain takes place as follows :

H_2 —— Intermediate carrier —— Flavoprotein —— Cytochromes —— O_2

|
NAD

2. Nitrate reduction

This has been seen in bacteria commonly called denitrifying bacteria (Micrococcus denitrificans). These bacteria use nitrate as the terminal acceptor of electrons instead of oxygen. Nitrate is reduced with molecular hydrogen. The bacteria possess type 1 hydrogenase enzyme. As NAD cannot be reduced directly, an ATP dependent reduction system is used. The reduction of nitrate is believed to be a process occurring in several steps as follows:

$$NO_2 \text{——} NO_2 \text{——} NO \text{——} NO_2 \text{——} N$$

3. Sulphate reduction

Two types of sulphate reduction have been noticed in bacteria. These are

Assimilatory sulphate reduction, and *Dissimilatory sulphate reduction.*

Assimilatory sulphate reduction

In this process, sulphate is reduced to sulphide which is used for the synthesis of sulphur containing amino acids like cysteine and methionine. This is called Assimilatory sulphate reduction because the sulphate is assimilated into the amino acids. The following steps are involved in Assimilatory sulphate reduction.

1. Activation of sulphate by ATP. In this process, inorganic sulphate reacts with ATP and is catalysed by the enzyme ATP sulphurylase resulting in the formation of APS.
2. Phosphorelation of APS. APS is phosphorelated with the help of ATP to produce a compound called PAPS (3 phospho adenosine, 5 phospho sulphonate).
3. Reduction of sulphate. The sulphate in PAPS is reduced to sulphite by the action of the enzyme PAPS reductase. This reduction involves the transfer of electrons from NADH to PAPA.
4. Reduction of sulphite. NADPH reduces sulphite to sulphide under the influence of the enzyme sulphite reductase.

$$SO3 — 3NADPH + 3H^+ \text{———} H_2S + N3NAD + + 3H_2O$$

The hydrogen sulphide released in the above reaction is immediately utilised for the formation of ammino acid cysteine. This reaction is catalysed by the enzyme sulphydrase.

Dissimilatory sulphate reduction

A few species of gram negative (*Desulfovibrio*), and gram positive (*Desulfotomaculum*) bacteria are capable of dissimilatory sulphate reduction. These bacteria are strictly anaerobic.

Dissimilatory sulphate reduction produces hydrogen sulphide which is found in the cytoplasm of bacteria such as *Desulphovibrio.* These bacteria can also use pyruvate, lactate etc. as substrates. *D.Desulfuricans* can also utilise molecular hydrogen for the reduction of sulphate as follows :

$$SO_4 — + 4H_2 — S — + 4H_2O$$

The first step in the reduction of the sulphate is the formation of APS. The reaction is similar to what is seen in the Assimilatory sulphate reduction. In the next step, APS is reduced to sulphite and AMP under the influence of the enzyme APS reductase. Further reduction of sulphide is brought about by the enzyme sulphite reductase. In the above process, Cyt c3 is an electron carrier.

4. Methanogenesis

In certain bacteria, oxidation of hydrogen takes place and carbon dioxide acts as a terminal electron acceptor. Reduction of carbon dioxide here results in the production of methane. Bacteria which reduce carbon dioxide to methane are called methanogenic bacteria or methanogenic bacteria or methanogens. Mathanogenic bacteria are found in the intestine of many animals and also in soil and in sewage tanks. Physiologically, they are strict anaerobes but Autotrophic, as they use only carbon dioxide as the source. A few examples of methanogenic bacteria are *Methanobacterium formicicum, M.Ruminaticum, M. Mobilis M.Propionicum etc.*